Cambridge Primary

Revise for Cambridge Primary Checkpoint

Science
Study Guide

Peter D. Riley

Series editor: Caroline Wood

HODDER
EDUCATION
AN HACHETTE UK COMPANY

Contents

Physics

Glossary 92

About this Study Guide

What's in this book?

Science is a fascinating subject! You have studied it for some years now. This book helps you to prepare for the Cambridge Primary Checkpoint Science tests at the end of Stage 6.

All the topics you have studied in Science are revised in this Study Guide. There are three chapters: Biology, Chemistry and Physics. Each chapter contains the following sections to help you with your revision.

Key facts This section gives you important information about the topic. You need to revise and learn this information as much as possible.

Key words This section contains words that may appear in the test. Use a dictionary to find their meaning or for words in bold in the **Key facts** section use the glossary on pages 92–95.

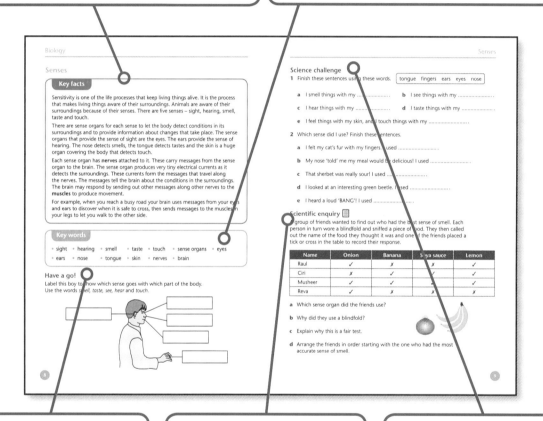

Have a go! This section contains practice questions.

Scientific enquiry This section contains questions that practise your scientific enquiry skills.

Science challenge Try these slightly more difficult questions to test your knowledge of the topic!

How much can you remember? Tests

At the end of each chapter there is a short practice test called **How much can you remember?** covering a variety of topics. The questions are written in a similar way to the Cambridge Primary Checkpoint tests so that you become familiar with the style of questions in the real tests.

Tips for revision

- Find somewhere quiet to revise. Sit at a table on a comfortable chair with the Study Guide and a pen or pencil. Have some paper ready for answering longer questions (these have this symbol: 📑).

- Turn to the section you wish to revise and read the **Key facts**, then move on to the **Key words**.

- Reading through the text is not always the best way to revise. It is better to look up the **Key words** in a dictionary and then re-read the **Key facts**. This time, when you reach a key word, think of its definition.

- Work through the sections in order, starting with **Have a go!**. Move on to **Science challenge** and finish with **Scientific enquiry**.

- Read each question carefully and record your answers as instructed.

- When you have finished a chapter, look through the pages again and try the short practice test to check your knowledge.

- If you find there are some parts of a topic you have difficulty remembering, read through the page again. A second look often makes the missing facts clearer in your mind.

Units of measure

Make sure you revise and understand these units of measure and their symbols.

Quantity	Unit	Symbol
length	kilometre	km
	metre	m
	centimetre	cm
	millimetre	mm
mass	kilogram	kg
	gram	g
time	hour	h
	minute	min
	second	s
volume and capacity	cubic centimetre	cm³
	litre	l
	millilitre	ml
temperature	degree Celsius	°C
force	newton	N
sound	decibels	dB

Biology

Life processes

Have a go!

1 What are the life processes shown in pictures A–D?

A .... B ....

C .... D ....

2 Draw lines to match each process with its definition.

A nutrition		**1** change position
B movement		**2** increase in size
C growth		**3** produce offspring
D reproduction		**4** acquire food and water

Science challenge

Group these things into *Alive*, *Once alive* and *Never alive*. It is more difficult than it looks!

teddy	plastic toy horse	pot plant
child	dried grass	pebble
pen	empty whelk shell	flower in a vase

Alive	Once alive	Never alive

Scientific enquiry

1 Tick (✓) to show which life processes these things carry out.

Life process	Hamster	Plant	Robot toy
nutrition			
movement			
growth			
reproduction			
respiration			
sensitivity			
excretion			

2 Which of the above are living things? ....

3 How do you know they are living things? ....

Senses

Key facts

Sensitivity is one of the life processes that keep living things alive. It is the process that makes living things aware of their surroundings. Animals are aware of their surroundings because of their senses. There are five senses – sight, hearing, smell, taste and touch.

There are sense organs for each sense to let the body detect conditions in its surroundings and to provide information about changes that take place. The sense organs that provide the sense of sight are the eyes. The ears provide the sense of hearing. The nose detects smells, the tongue detects tastes and the skin is a huge organ covering the body that detects touch.

Each sense organ has **nerves** attached to it. These carry messages from the sense organ to the brain. The sense organ produces very tiny electrical currents as it detects the surroundings. These currents form the messages that travel along the nerves. The messages tell the brain about the conditions in the surroundings. The brain may respond by sending out other messages along other nerves to the **muscles** to produce movement.

For example, when you reach a busy road your brain uses messages from your eyes and ears to discover when it is safe to cross, then sends messages to the muscles in your legs to let you walk to the other side.

Key words

- sight - hearing - smell - taste - touch - sense organs - eyes
- ears - nose - tongue - skin - nerves - brain

Have a go!

Label this boy to show which sense goes with which part of the body.
Use the words *smell, taste, see, hear* and *touch*.

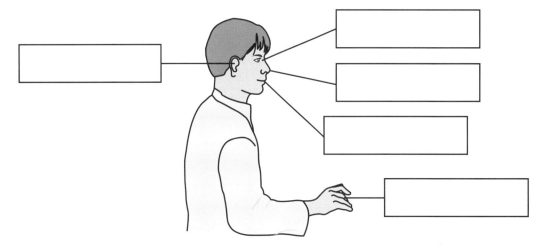

Science challenge

1 Finish these sentences using these words.

> tongue   fingers   ears   eyes   nose

a I smell things with my ....   **b** I see things with my ....

c I hear things with my ....   **d** I taste things with my ....

e I feel things with my skin, and I touch things with my ....

2 Which sense did I use? Finish these sentences.

a I felt my cat's fur with my fingers. I used ....

b My nose 'told' me my meal would be delicious! I used ....

c That sherbet was really sour! I used ....

d I looked at an interesting green beetle. I used ....

e I heard a loud 'BANG'! I used ....

Scientific enquiry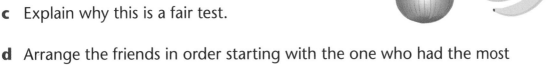

A group of friends wanted to find out who had the best sense of smell. Each person in turn wore a blindfold and sniffed a piece of food. They then called out the name of the food they thought it was and one of the friends placed a tick (✓) or cross (✗) in the table to record their response.

Name	Onion	Banana	Soya sauce	Lemon
Raul	✓	✗	✗	✓
Ciri	✗	✓	✓	✓
Musheer	✓	✓	✓	✓
Reva	✓	✗	✗	✗

a Which sense organ did the friends use?

b Why did they use a blindfold?

c Explain why this is a fair test.

d Arrange the friends in order starting with the one who had the most accurate sense of smell.

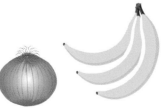

Classifying living things

Key facts

Scientists sort living things into groups by looking at how they are similar to and different from each other. For example, they put living things that have roots, stems and leaves in the plant group. They put living things that do not have these features in the animal group.

They can then sort the plants and animals into smaller groups by looking again for similarities among them. For example, we can divide animals into different groups according to whether they have fur, feathers, scales, legs or wings.

The sorting of living things into groups is called **classification**. Scientists use keys to identify living things.

Key words

- classification
- key

Have a go!

1 What groups could you sort these organisms into?

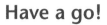

....

....

2 Use the key to identify animals A and B.

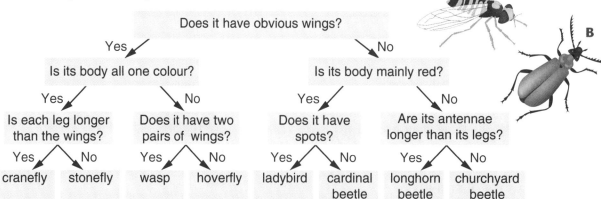

Animal A is a ....   Animal B is a ....

Science challenge

1 Look at the key in question 2 on the opposite page and then make keys to classify the animals below: Write a question to separate the first two animals from the second two, and then a question to separate each pair. One has been done for you.

a slug, snail, beetle, butterfly

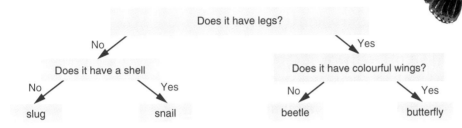

Does it have legs?

No → Yes →

Does it have a shell   Does it have colourful wings?

No →   Yes →   No →   Yes →

slug   snail   beetle   butterfly

b zebra, tiger, tortoise, crocodile

c hawk, finch, snake, lizard

2 Scientists also use Venn diagrams to sort living things into sets. Draw a line from each animal to the correct part of the diagram.

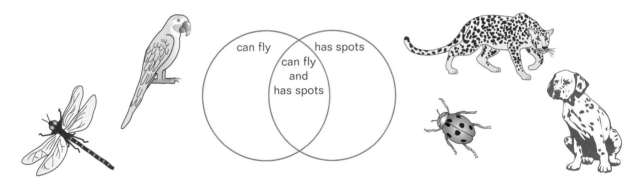

can fly   has spots

can fly and has spots

3 Use the Venn diagram to sort these plants into groups. Think of the three questions you will ask first.

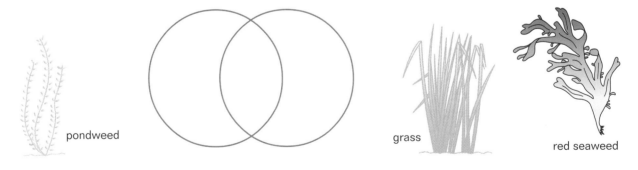

pondweed   grass   red seaweed

Scientific enquiry

Make a survey of pets your classmates keep at home. Arrange a classification for them.

Parts of a flowering plant

Key facts

The four main parts of a flowering plant are the roots, the stem, the leaves and the flowers.

The roots hold the plant firmly in the ground and take up water and **minerals** from the soil.

The stem holds up the leaves and flowers and carries water and minerals from the roots to them. It carries food that the leaves make to all parts of the plant.

The leaves make food from sunlight, water and carbon dioxide in the air.

The flower makes seeds that can grow into new plants.

The roots need to be healthy to take in water and minerals. The leaves need to be healthy to make food and the stems need to be healthy to transport the water, minerals and food around the plant. If all these parts are healthy then the plant is healthy and can produce flowers and seeds. You can tell if a plant is unhealthy. A plant that does not get enough light will have pale or yellow leaves. A plant not getting enough water will have brown leaves that may curl up.

Key words

- root
- stem
- leaf
- flower
- minerals
- carbon dioxide
- seeds

Have a go!

Label the parts of the flowering plant. What job does each part do?

A .... My job is ....

....

B .... My job is ....

....

C .... My job is ....

....

D .... My job is ....

....

Science challenge

1 Circle the best words to complete the sentences.

a The root / stem / leaf helps to hold the plant steady in the soil. It also helps the plant to take up sunlight / water / beetles from the soil.

b The root transports water / food / sunlight straight to the leaf / stem / flower.

c The stem transports water / food / minerals from the leaf to the rest of the plant.

d A plant produces seeds in its leaves / roots / flowers when all parts of the plant are healthy / unhealthy / absent.

e The leaves / roots / flower make the water / minerals / food, which reaches the root after it has passed through the stem / flower / leaf.

2 a Describe the path water takes from the soil to the leaf.

....

....

b Describe the path the food takes from the leaf to the flower.

....

....

Scientific enquiry

1 What would happen to a plant that has unhealthy roots?

2 The pictures show the leaves of three plants.

a Which plant is a healthy plant? ....

b Which plant needs more light? ....

c Which plant needs more water? ....

Plant reproduction

Key facts

The flower contains the reproductive organs of the plant, which make and receive pollen. A flower receives pollen in a process called pollination. Insects pollinate many flowers; these flowers have large bright petals, and also produce scents and nectar, to attract insects.

Inside the petals are the male reproductive organs, called stamens, which have two parts – an anther, which produces pollen, and a filament, which is a stalk that supports the anther.

At the centre of the flower is the female reproductive organ called the carpel. It is divided into three parts, called the ovary (which contains ovules), the stigma and the style. The style connects the stigma to the ovary and the stigma has a sticky surface to catch pollen.

Insects pick up pollen from the anthers and take them to the stigmas of other flowers of the same type of plant. Each pollen grain then grows a tube down to an ovule and part of it joins with it in a process called **fertilisation**. The fertilised ovule becomes a seed and the ovary then becomes a fruit to take the seeds away from the plant.

Key words

- pollen
- pollination
- nectar
- fertilisation
- stamen
- anther
- filament
- ovary
- ovules
- carpel
- stigma
- style

Have a go!

Match the labels below to the correct parts of this flower.

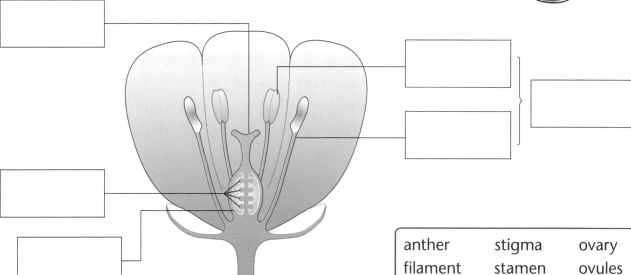

anther	stigma	ovary
filament	stamen	ovules

Science challenge

1 Use the words in the box to answer the questions.

| stigma petal filament anther |
| ovary stamen carpel |

   a Which part of the flower is brightly coloured and scented

      to attract insects? ....

   b Which part makes the pollen? ....

   c Which part contains the ovules? ....

   d Which part collects pollen? ....

   e Which part has a sticky surface? ....

   f Which part holds up the anther? ....

   g Which part is the male reproductive organ? ....

   h Which part is the female reproductive organ? ....

2 Circle the best words to complete the sentences.

   a The ovary of the plant contains the ovaries / ovules / ovoids. After fertilisation / respiration / excretion, the ovules develop into leaves / roots / seeds.

   b The stamen is made from a leaf / anther / root and a stem / petal / filament. It is found between the ovules / petals / style and the ovary / stem / leaf.

   c The anther / ovary / stigma makes pollen and spiders / insects / snails transport it in a process called fertilisation / pollination / propagation.

   d When a pollen grain lands on the anther / filament / stigma it sticks to it and grows a root / tube / stamen down to a seed / ovule / leaf in the anther / ovary / petal where the process of fertilisation / pollination / propagation takes place.

Scientific enquiry

A plant has a flower in which the anthers produce pollen that falls off onto the stigma and the flower pollinates itself. How could you make sure that the flower only receives pollen from the flowers of other plants?

....

....

....

Life cycle of flowering plants

Key facts

The stages in a plant's life make up its life cycle. Below is the life cycle of an apple tree.

The seed **germinates**. A root grows first, and then a shoot.

The apple contains seeds. These spread when animals eat the apple – and the cycle starts again.

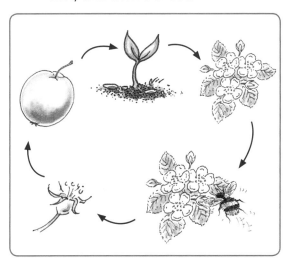

The plant grows larger and produces flowers. The flowers contain pollen.

Once **fertilised**, the ovary starts to develop into a fruit – a tiny apple.

Insects, such as bees, visit the flowers and pollinate them.

The apple disperses its seeds by letting animals eat the fruits while the seeds pass through the animal unharmed. Other plants disperse their seeds by letting them be hooked onto animals, carried by the wind or water or having their fruits explode into the surrounding environment.

Key words

- root
- germination
- shoot
- pollination
- fertilisation
- disperse

Have a go!

Write a number in each box to show the life cycle of a plant in the correct order.

a germination ☐   poppy seed ☐   poppy plant grows ☐

  seeds produced ☐   fertilisation ☐   pollination ☐

b pea seed ☐   pea plant grows ☐   seeds produced ☐

  fertilisation ☐   pollination ☐   germination ☐

c seeds produced ☐   fertilisation ☐   plant grows ☐

  germination ☐   pollination ☐   sunflower seed ☐

Science challenge

1 Use the words in the box to show how each seed is dispersed.

birds   fur   wind   water   explosion

a coconut

b strawberry

c dandelion

d Himalayan balsam

e cleavers

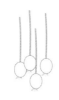

f mistletoe

g elderberries

h sycamore

2 Which of these sentences are true and which are false?
Write 'T' or 'F' in each box.

a Seeds germinate before they are dispersed. ☐

b Fertilisation occurs before the seeds are dispersed. ☐

c Seeds with tiny hooks on them are dispersed by the wind. ☐

d Birds can disperse the seeds of juicy fruits. ☐

e Some seed pods explode when they dry out. ☐

f Seeds with tiny hooks on them are dispersed by animals such as snakes that have smooth skin. ☐

Scientific enquiry 📄

The longer a seed stays in the air the further it can be dispersed.

a Name two seeds that are dispersed by the wind.

b Describe a fair test you could carry out to find out if one stays in the air longer than the other.

c Would you carry out the test more than once?

d Explain your answer to part **c**.

Plant growth

Key facts

Green plants need certain conditions to grow healthily.

They need enough water. If a plant lacks water it **wilts**, then its leaves become brown as the plant dries up.

Plants need sunlight because the leaves use some of the **energy** in sunlight to make food. They make the food from carbon dioxide in the air and water from the soil in a process called photosynthesis. If a plant lacks light it may grow tall and thin to try and find it while its leaves lose their green colour and may even turn yellow.

Temperature affects the growth of a plant. A plant grows faster in a warmer temperature than a cool temperature.

When **germination** takes place in a seed a plant grows out of it. Seeds need water and warmth to help them germinate.

All plants also need **minerals** for healthy growth. The minerals are in the soil but they also **dissolve** in water and pass into the plant in water the roots take up. Farmers and gardeners add **fertilisers** and **manure** to soil to provide extra minerals for growing plants.

Key words

- wilt
- sunlight
- carbon dioxide
- photosynthesis
- germination
- minerals
- fertilisers
- manure
- temperature

Have a go!

1 What materials does a plant need to make food?

....

2 What else does a plant need for photosynthesis to take place?

....

3 Why may a plant die if its leaves are removed?

....

4 What is wrong with a plant that has a drooping stem and floppy leaves?

....

Science challenge

Re-read the **Key facts** box opposite about growing conditions. In each box below draw and colour a plant to match the conditions described below them. Use the pictures to help you.

a	b	c	d

This plant grew on a bright, sunny shelf. It has plenty of water.

This plant grew on a bright, sunny shelf. It has not had enough water.

This plant grew in a dark cupboard. It has had plenty of water.

This plant grew in a dark cupboard. It has not had enough water.

Scientific enquiry

Ahmed and Su Lin had six geraniums that were the same height.

They took half the leaves off three plants. They cared for all the plants in the same way for two weeks and then measured their heights. The table shows their results.

Plant	Height in cm
1 – half the leaves removed	14
2 – half the leaves removed	13
3 – half the leaves removed	11
4 – no leaves removed	20
5 – no leaves removed	19
6 – no leaves removed	20

a Which plant grew least? ....

b Which group of plants varied the most in height? ....

c Why could Ahmed and Su Lin say that the difference in height was due to the number of leaves?

d Describe an experiment to see if fertiliser makes plants grow larger.

Plant experiments

> ### Key facts
>
> Scientists can perform experiments on plants to see what **factors** affect their **germination** and growth. For germination, scientists can investigate the seeds to see if they need light, warmth or water for them to sprout. Once the seeds have germinated, scientists can investigate the seedlings to see how light, warmth and water affect their growth.
>
> For each experiment, scientists change the one factor they are investigating and keep all the others they are not investigating the same. This makes the experiment a **fair test**. For example, if scientists were investigating seeds to see if they needed light they would give them the same amount of water and keep them at the same **temperature**. If scientists were investigating seedlings to see how much water they needed for growth, they would give each seedling a different amount of water but keep them at the same temperature and give them the same amount of light.
>
> In plant experiments, scientists make measurements by using rulers, measuring cylinders and thermometers and record the results in tables and present them in bar charts and line graphs.

> ### Key words
>
> • factors • fair test • germination • growth • light • warmth • water

Have a go!

Marvin and Sonia set up two beakers like the one shown in the diagram.

They put one beaker in a warm place at 25 °C and one in the fridge at 4 °C.

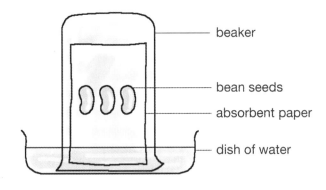

a Write down two factors that they kept the same.

....

b Write down one factor that they changed. ....

c Which beans do you think germinated first? ....

d Explain your prediction. ....

....

Science challenge

Alex did an experiment to find out how much water plants needed to grow. He took four geranium plants, which were all 10 cm tall. He watered the plants every day but gave each plant a different amount of water. He placed all the plants in the same area to grow for two weeks and then measured their heights. The table shows the results.

Amount of water in cm³	Height after two weeks in cm
0	4 (plant had died)
5	14
20	15
100	5 (plant had died)

Growth of plants with different amounts of water

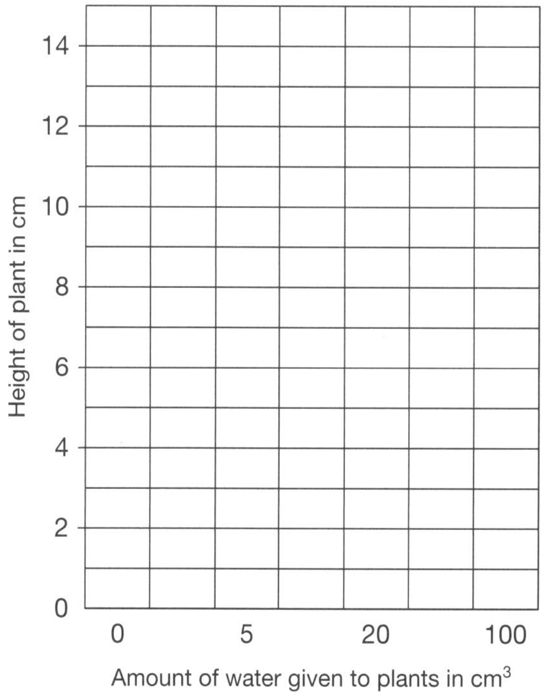

Height of plant in cm

14
12
10
8
6
4
2
0

0   5   20   100

Amount of water given to plants in cm³

a Complete the bar chart to show the results.

b What do you think Alex used to measure the

water he gave to the plants? ....

c What do you think he used to measure the

heights of the plants? ....

d What conclusion can you draw from the evidence of this experiment? Tick (✓) one box.

Plants need water to grow, but not too much. ☐

The more water you give a plant, the better it grows. ☐

20 cm³ is too much water to give a plant each day. ☐

Scientific enquiry

Zoe put a celery stem in red food colouring. The picture shows what it looks like after 9 hours.

a What does the celery look like after 9 hours? ....

b Why does it look like this? ....

....

c Zoe leaves a white flower in blue food colouring for 9 hours.

What would she see happening? ....

Habitats

> ### Key facts
>
> A **habitat** is a place where plants and animals live. There are many different habitats. Each habitat has its own special features. For example, a woodland has a soil that can provide a large number of plants with water and **minerals**. The trees that grow there provide shade and shelter for both plants and animals.
>
> The organisms that live in a habitat are **adapted** to live in the habitat. For example, some plants, such as ferns, are adapted to live in the shade of the trees, and squirrels are adapted to climb trees and feed on their fruits.
>
> Many habitats are aquatic habitats. We can divide them into freshwater habitats, such as rivers and ponds, and sea-water habitats, such as the seashore or the open sea itself.

> ### Key words
>
> * habitat
> * organisms
> * adaptation
> * freshwater
> * woodland

Have a go!

1 a Which of these organisms would you find in a wood? Circle them.

b Name the organisms you have not circled. Write down the habitat for each.

....

....

2 Draw lines to match each list of organisms to the correct habitat.

A	vulture, scorpion, sand rat
B	polar bear, snowy owl, reindeer
C	monkey, parrot, tree frog
D	whale, tuna, squid

1	open sea
2	rainforest
3	area near North Pole
4	desert

Science challenge

Match each organism to a set of adaptations in the table. Then match the habitats to the organisms.

Organisms:
seal
fennec fox
limpet
hippopotamus
bush baby

Habitats:
swamp
seashore
rainforest
open sea
desert

	Adaptations	Organism	Habitat
a	large eyes for seeing in dark conditions, small paws with nimble fingers for picking and peeling fruit		
b	thick layer of blubber for keeping warm, big eyes to see in gloomy water, streamlined shape and flippers		
c	huge ears to help in cooling the body, yellow or light brown colour for camouflage		
d	thick leather skin to protect against the heat, eyes and nostrils on top of head so they are above water surface while body is in the water		
e	has one foot that is like a sucker to grip a rock and a shell to keep the body from drying out when out of water		

Scientific enquiry

The picture shows a choice chamber. One half has water in it, which makes the air damp, and the other half has a chemical called a drying agent, which makes the air dry. Woodlice are grey shrimp-like organisms that live on woodland floors. How could you use the choice chamber to find out if woodlice prefer damp or dry air?

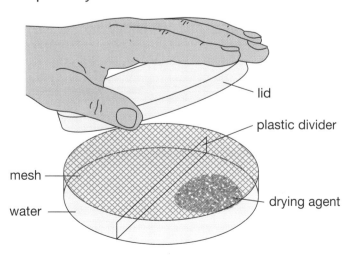

23

Food chains

Key facts

All living things need food to give them **energy**. Plants make food in their leaves using sunlight. They are known as producers of food. Animals cannot make their own food so must obtain it from plants or other animals. They are known as **consumers** of food.

The way that producers and consumers are related to each other is shown in a **food chain**. The energy passes from the producer through the consumers and its path is shown by arrows that go from the eaten to the eater:

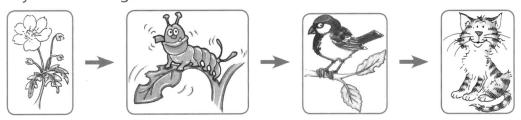

The caterpillar is a consumer known as a **herbivore** because it eats plants. It is also prey to the sparrow, which is a **carnivore** and a predator on the caterpillar. A carnivore eats meat. The sparrow is also prey to the cat, which is its predator. As the cat is at the end of the food chain and not prey to any other animal it is called the top carnivore.

Key words

- food chain
- energy
- producer
- consumer
- herbivore
- prey
- carnivore
- predator

Have a go!

Here are five food chains found in a forest but they are in the wrong order.
Rewrite them in the correct order in the boxes on the right.

a slug → leaf → thrush   [ ] → [ ] → [ ]

b fox → grass → rabbit   [ ] → [ ] → [ ]

c deer → plants → wolf   [ ] → [ ] → [ ]

d plant → blackbird → caterpillar   [ ] → [ ] → [ ]

e owl → nuts → mouse   [ ] → [ ] → [ ]

Science challenge

1 Sort these creatures into the correct part of the table below.

cat	sheep	owl	fox
rabbit	goat	frog	
slug	greenfly	cow	spider

Herbivores (plant eaters)	Carnivores (meat eaters)

2 a Circle the top carnivores.

white shark   tiger   mouse   polar bear   frog   antelope

camel   penguin   giraffe   squirrel   lion   orca (killer whale)

zebra   ox   rabbit   eagle   alligator   chicken   horse

b Describe how you chose which animals were the top carnivores.

....

....

....

Scientific enquiry

What do you predict would happen in the food chain below from a river if a pollutant in the water killed off all the snails? Explain your answer.

water plants → snails → fish

....

....

Ways humans have affected the environment

Key facts

Our environment is our surroundings. It is made up of the atmosphere, the **habitats** of the world and the water that runs through them or forms them. Over the years the human population has affected the environment in the following ways.

Power stations using **fossil fuels** have polluted the atmosphere with **greenhouse gases**. Exhaust fumes from millions of cars and trucks produce these gases too. Power stations also produce a gas that makes acid rain. People have destroyed habitats to get at metal ores in the ground, then heated them to release the metals, causing air and water pollution. When factories work the metals with other **materials** such as plastics to make a wide range of goods, more air and water pollution occurs.

Even when people no longer need the goods they can cause more environmental damage by leaving them as litter in the countryside. Broken glass can cut deer and other large animals, open bottles can trap mice and other small animals, and plastic bags can cover plants and prevent them receiving the light they need to make food.

Key words

- atmosphere
- power station
- exhaust fumes
- fossil fuels
- ores
- greenhouse gases
- acid rain
- water pollution
- goods

Have a go!

1 Draw lines to match each human activity with the environmental damage.

Activity

| **A** power station making electricity |
| **B** taking metal ores |
| **C** leaving litter |

Environmental damage

| **1** large animals cut by glass |
| **2** greenhouse gases |
| **3** habitat destruction |

2 How can litter be a danger to woodland mice? ....

....

Science challenge

Dust and soot particles in the air can settle on plants and reduce the amount of light reaching the leaves.

a State how this could affect the plant. ....

....

b Describe how the dust and soot could affect caterpillars that feed on the

leaves. ....

....

c A caterpillar makes a pupa when it is fully grown and healthy. Inside the

pupa it forms a butterfly. How would the dust and soot affect the numbers

of butterflies seen in the habitat? ....

....

d How could you compare the amount of dust and soot falling in two areas

in your environment using white card and a magnifying glass?

....

....

Scientific enquiry

A scientist suspected that hot water was polluting a river and recorded the water temperature at seven places, called *stations*, along it. The table shows her results.

a At which station did water pollution

probably occur? ....

b What happened to the water as it flowed

away? ....

Station	Temperature in °C	Fish present
1	12	
2	12	
3	30	
4	25	
5	20	
6	15	
7	12	

c The scientist knew that fish A could not live in water above 12 °C, fish B could not live in water above 20 °C and fish C could not live in water above 25 °C.

Which fish did she predict would be present at each of the stations? Write their letters in the third column of the table.

Caring for the environment

Key facts

In many countries people are trying to take better care of the environment.

They try to use less electricity so power stations need to produce less, so reducing the production of **greenhouse gases**. People recycle **materials**. By recycling metals they reduce the need to destroy **habitats** to find metal ores. We use trees to make paper and cardboard so by recycling these items we need to cut down fewer trees. We make glass from molten sand and need large amounts of **fuel** to produce the heat. By recycling glass, we need less fuel and also reduce habitat damage. Factories use lots of chemicals to make goods and can release the chemicals into rivers and the sea and cause pollution. In many places factory owners make sure that they safely store the chemicals after use so they cannot cause pollution.

In the past, people destroyed habitats everywhere to get at raw materials but today governments have made many habitats into nature reserves to protect them from destruction. Where some habitats have been destroyed people are restoring them. As plants and animals depend on their habitats for their survival, **conserving** and restoring habitats save them from **extinction**.

The golden lion tamarin lives in the rainforest of Brazil and might become extinct because of rainforest destruction.

Key words

- recycle • fuel • raw materials • nature reserve • conservation • extinction

Have a go!

Fill in the missing words. Use the words in the box below.

| fuel | electricity | recycle | greenhouse gases | climate change | habitat |

a When we use less .... power stations produce smaller

amounts of ....

b Greenhouse gases may be a cause of ....

c When we .... materials we use less ....

and reduce .... damage.

Science challenge

1 Re-read the information in the **Key facts** sections on pages 26 and 28 and complete this table.

Humans' negative effects on the environment	Humans' positive effects on the environment

2 Name four materials in household rubbish that we can recycle.
Give an example of an item made from each material.

Material

....

....

....

....

Item

....

....

....

....

Scientific enquiry

Scientists have to use their imaginations to design things that they might need in an experiment. Imagine you needed a small vehicle to roll down a ramp. In the box below, draw a four-wheeled vehicle that you could make from items of household rubbish.

Major organs of the body

Key facts

The body is divided into the head, the **torso** (with the upper part being the **thorax** and the lower part being the **abdomen**) and the **limbs**, which are attached to the torso.

The major organ in the head is the brain. This receives information from the sense organs – the eyes, ears, tongue, nose and skin – and sends out messages along **nerves** to make the **muscles** move.

In the chest are a pair of lungs that take in oxygen from the air and release carbon dioxide in a process called breathing.

Between the lungs is the heart, which pumps blood around the body in tubes called blood vessels. The blood carries oxygen and digested food to all parts of the body and removes waste from them.

In the upper abdomen are the stomach and the liver that, along with the intestines lower down, break down the food in a process called digestion. At the back of the upper abdomen are a pair of kidneys, which act like **filters** and remove waste from the blood. The kidneys excrete the waste in a **liquid** called urine.

Key words

- head
- torso
- limbs
- thorax
- abdomen
- brain
- lungs
- breathing
- heart
- stomach
- liver
- intestines
- kidneys

Have a go!

1 Use the **Key facts** above to label the organs of this body.

| brain | heart | lung | stomach |
| kidney | intestine | liver | |

2 Systems are groups of organs that work together.

Which three organs in the diagram form part of the digestive system?

....

....

....

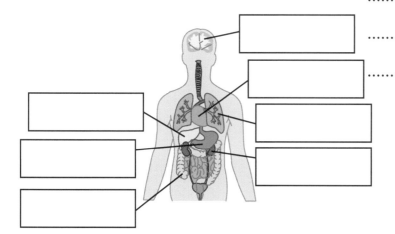

Science challenge

1 Answer these questions about the picture.

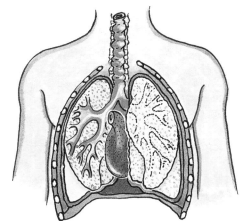

a Which part of the body is this – the thorax or abdomen?

....

b On the diagram write 'A' next to the organ that pumps blood around the body.

c On the diagram write 'B' next to the organ that takes in oxygen from the air.

d To make this diagram some of the ribs have been cut away but you can still see parts of them. Write 'C' next to the cut-away part of a rib.

2 The circulatory system carries blood to all parts of the body. Name three

things that travel in the blood.

....

Scientific enquiry

Take your pulse by placing your fingers as shown in the picture.

a Feel for a throbbing under the skin. This is your beating pulse.

The beating heart causes it. How many times does your pulse

beat in a minute when you are sitting down? ....

b Describe how you think your pulse will change after you have walked

around for a minute. ....

c Test your prediction and explain what you find.

....

....

....

....

....

Bones and muscles

Key facts

Some animals, such as humans, have a skeleton of bones inside their body. These bones grow as the body grows. They support and protect parts of the body and help us to move.

The power of movement comes from the **muscles** that are attached to the bones. A muscle can bring about movement when it contracts or gets shorter. The contracting muscles pull on a bone and can make it move. Muscles cannot push and they must be helped to relax and stretch again by the pulling action of another muscle. This means that muscles are arranged in pairs so that when one muscle contracts and moves a bone another muscle relaxes, stretches and is then ready to contract and move the bone in the opposite direction when needed.

The biceps and triceps are a pair of muscles that move the bones in the lower arm, as the pictures show.

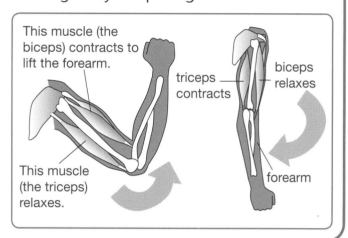

This muscle (the biceps) contracts to lift the forearm.

This muscle (the triceps) relaxes.

triceps contracts

biceps relaxes

forearm

Have a go!

1 Match the correct bones to the descriptions.

| ribs |
| pelvis |
| skull |
| spine |

a bony case to protect your brain

the stack of bones that runs down your back

a large bone that connects the legs to the rest of the skeleton and protects some of the lower organs inside the body

the bones that make a cage to protect your heart and lungs

Key words

- skeleton
- bone
- muscle
- contract
- relax

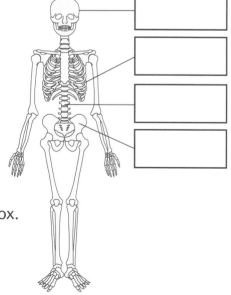

2 Label the parts of the skeleton with the words in this box.

ribs   pelvis   skull   spine

Science challenge

Draw lines from the following descriptions to the correct places on the picture of the leg. There are six places, each with a red dot on it.

the muscle that contracts to lift the lower leg

the muscle that relaxes when the lower leg is lowered to push down on something

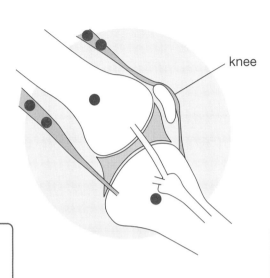

knee

a bone in the upper leg

a bone in the lower leg

the muscle that relaxes when the lower leg is lifted

the muscle that contracts when the lower leg is lowered to push down on something

Scientific enquiry

1 Straighten your right arm in front of you. Spread the fingers of your left hand and dig them into the muscle at the front of your upper right arm (your biceps) then move your right hand towards your face and feel how the muscle changes with your fingers.

Tick (✓) two boxes.

the muscle shortens ☐

the muscle lengthens ☐

the muscle becomes softer ☐

the muscle becomes harder ☐

2 Look at the picture of the skeleton on the opposite page and hold out your right hand with the palm up. Feel the bones in your lower right arm with the fingers and thumb of your left hand, then turn over your right hand. What happens to the bones?

....

....

....

....

Food

Key facts

A nutrient is a **substance** in a food that provides us with **material** for growth, **energy** or good health. The nutrient groups are **proteins, fats, carbohydrates**, vitamins, **minerals** and **fibre**. We also need water.

Proteins help us to grow, and repair cuts and bruises. Cheese, meat, fish, milk and nuts provide protein. Fats give us lots of energy and the body can store them under our skin to keep us warm. Butter, cheese, nuts and cooking oils provide fats. Carbohydrates can supply energy that the body can use quickly, and can also be used as an energy store in the body. Pasta, rice and cereals provide carbohydrates. Vitamins have many uses in keeping the body healthy, and fresh fruit and vegetables provide them. We need minerals for health and growth and can find them in milk, cheese, eggs, fruit and vegetables. Fibre gives the food bulk, which the **muscles** in the digestive system can push against to keep the food moving as it is being digested. It prevents constipation. We can find fibre in bread, fruit and vegetables. We should eat sugar and fats in small amounts as sugar can cause the teeth to rot and fats can build up in the body, making it very heavy and difficult to move and cause the heart to work harder to pump the blood.

Have a go!

1 a Match the food to the food group. Draw a line from each food to the correct name.

| protein | fats | carbohydrates | fibre |

Key words

- proteins
- fats
- carbohydrates
- vitamins
- minerals
- fibre
- constipation
- sugar

b Which of these foods contains both protein and fat?

....

2 Circle the foods that may contain large amounts of sugar.

pasta   toffee   cake   milk   lemonade   carrot   biscuit   bread

Science challenge

1 Which of these sentences are true and which are false? Write 'T' or 'F' in each box.

a If you eat all the different nutrients you do not need water. ☐

b The body can store fats. ☐

c Fibre is found in cheese. ☐

d We need vitamins to keep the body healthy. ☐

e The body can use the energy in pasta quickly. ☐

f Too much sugar and fat in the diet can damage your health. ☐

2 a If you had a meal of bread and cheese what nutrients would you be eating?

b Describe how you would be helping your body.

Scientific enquiry

A scientist performed some experiments on 100 grams of different foods to find out how much protein, fat and carbohydrate they contained. The table shows his results.

a Which food is richest in carbohydrate?

....

b Which food has almost equal amounts of the three nutrients?

....

Food	Protein in g	Fat in g	Carbohydrate in g
chicken	20	6	0
lentil	23	0	53
peanut	28	49	8
milk	3	3	4
rice	6	1	86
egg	12	10	0
fish	16	18	0
lamb	15	30	0

c How much protein is there in 200 g of chicken? ....

d What is the difference in the fat content of 100 g of peanuts and 100 g of lamb?

....

e Which food provides protein and carbohydrate but no fat to a meal?

....

Keeping healthy

> ### Key facts 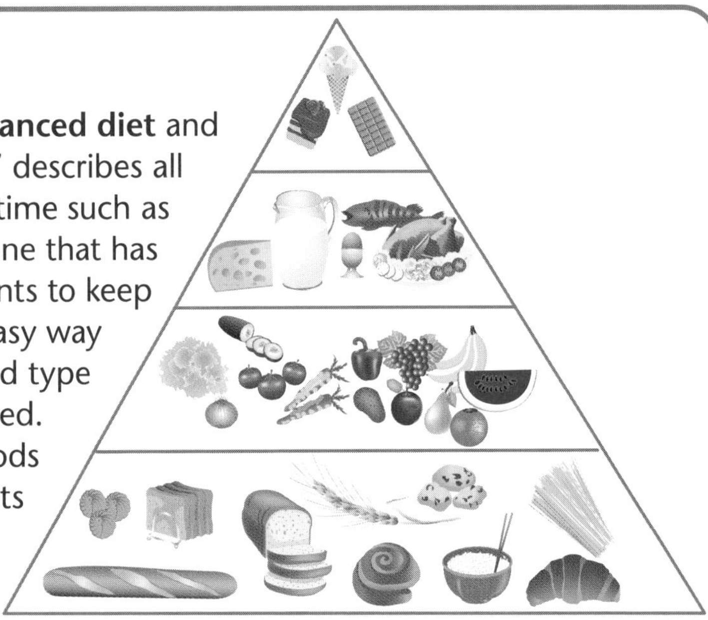
>
> People can stay well if they eat a **balanced diet** and take regular exercise. The word 'diet' describes all the food you eat during a period of time such as a day or a week. A balanced diet is one that has all the nutrients in the correct amounts to keep you healthy. The picture shows an easy way to remember how much of each food type you can eat to keep your diet balanced. You can eat large amounts of the foods at the bottom but only small amounts of those at the top.
>
> Taking exercise every day helps all the organs of the body to work well. Exercise uses up **energy** and helps stop fat building up in the body, which can damage the heart. Exercise also makes the heart and **muscles** stronger and helps the joints between the bones move more easily.
>
> When people are ill doctors may prescribe them medicines to make them better. The medicines contain drugs that are chemicals that attack the disease and also ease the pain while the body is becoming well again.

Have a go!

Key words

- balanced diet
- exercise
- medicines
- drugs

1 Which foods should we only eat small amounts of in our diet?

....

2 Which foods can we eat in the largest amounts in our diet?

....

3 Tick (✓) the healthier option out of each pair of meals.

a chicken and vegetable stir fry + orange juice ☐ *or* cheese pasty and chips + regular lemonade ☐

b beans on toast + glass of milk ☐ *or* doughnut + chocolate bar + high energy drink ☐

c burger and fries + limeade ☐ *or* carrot soup with a bread roll + apple juice ☐

Science challenge

1 True or false? Write 'T' or 'F' in each box.

a Exercise damages the heart. ☐

b Exercise makes muscles stronger. ☐

c Eating too much fatty food can damage your health. ☐

d You should eat as much sugar as possible as it gives you lots of energy. ☐

e You should eat only small amounts of nuts and chocolates and larger amounts of fruit and vegetables. ☐

2 Fill in the missing words from the box below to complete the sentences.

| growth   chicken   balanced   cheese   healthy   chocolate   nutrients   fish |

a You should aim to eat a .... diet as this will give you all the .... you need for .... and to stay ....

b You should eat small amounts of .... and .... and larger amounts of .... and ....

3 Circle the correct words to use in the sentences.

a Exercise stops the body building up fat / protein / vitamins and stops damage to the heart / brain / kidney. It makes the muscles shorter / stronger / weaker and helps the joints / intestine / nutrients between the teeth / bones / lungs move more easily.

b If someone is healthy / ill / taking exercise a doctor may prescribe them a medicine / carrot / drug that contains a cheese / medicine / drug to make them ill / worse / better.

Scientific enquiry

Plan an investigation to find out how five minutes of different sports affect the beating of the heart. In your plan write down:

a the sports you would investigate

b the factors you would control

c what you think you will find out and a reason for your prediction

d what apparatus you would use

e what measurements you would make.

How much can you remember? Test 1

1 There are seven life processes. Here are three of them.

   nutrition   excretion   sensitivity

   (a) What are the other four?

   ....   [4]

   (b) What does **nutrition** mean? ....   [1]

   (c) What does **excretion** mean? ....   [1]

2 Here are eight animals.

   (a) Write down three ways that you
   could sort them into groups.

   ....

   ....

   ....   [3]

   (b) What word describes sorting living things into groups?

   ....   [1]

3 Answer these questions about the parts of a plant.

   (a) Name two tasks the roots perform.

   ....   [2]

   (b) Name two tasks the stem performs.

   ....   [2]

4 Complete these sentences about the parts of a flower.

   Choose words from the list.

   filaments   ovary   rough   petals   sticky   anthers   ovule   smooth

   (a) The parts that attract insects to the flower are called ....   [1]

   (b) The parts of the flowers that produce pollen are called ....   [1]

   (c) The surface of the stigma is .... to catch pollen.   [1]

   (d) The pollen grain on the stigma grows a pollen tube down to an ....   [1]

5 Nico is growing three plants at different temperatures.

He checks the height of each one is the same. He puts one in a cold place, one in a warm place and one in a very warm place.

(a) What must he do to make his test fair? ....

.... [2]

Nico measures the height of each plant every three days.

This table shows his results.

Plant	Day 0 Height in cm	Day 3 Height in cm	Day 6 Height in cm	Day 9 Height in cm
in cold place	6	7	8	9
in warm place	6	9	12	15
in very warm place	6	10	14	18

(b) How much did the plant in the cold place grow in 6 days? .... [1]

(c) What is the difference in height after 9 days between the plant in the very

warm place and the plant in the warm place? .... [1]

(d) What conclusion can Nico draw from these results? ....

.... [1]

(e) Describe how Nico could make his investigation more reliable. ....

.... [1]

6 The diagram shows the outline of a human body.

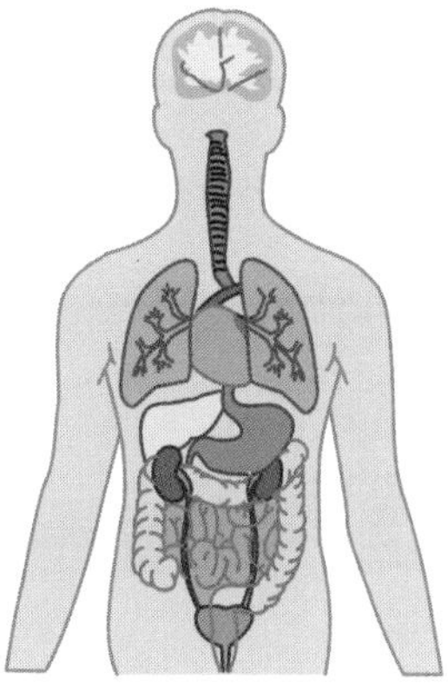

(a) Add labels to the diagram to show the brain, lungs, heart and stomach. [4]

(b) What is the purpose of the heart?

.... [1]

(c) In which organ does digestion take place? [1]

....

Chemistry

Materials

The word 'material' can mean a piece of cloth. However, in Science we use the word to describe what an object is made from. For example, scientists consider wood, water and even air to be a material. Every material has a number of features. Scientists call these features the **properties** of the material.

We can describe materials in many different ways according to their properties. Here is a list of the common properties of materials.

soft	flexible	rigid	rough	smooth	dull
shiny	transparent	opaque	absorbent	waterproof	

A material may not have all these properties. For example, think of the metal in a coin. The metal is hard, rigid, smooth, shiny, **opaque** and **waterproof**. All of these properties make metal very useful for making coins. Just think what coins would be like if we made them from a material that was soft, flexible, rough, dull, **transparent** and **absorbent**!

When someone is making an object, they select the materials that have useful properties that will help the object perform its task. An umbrella, for example, is made from a frame of metal rods because metal is rigid and flexible, and waterproof fabric is placed over the rods to provide shelter from the rain.

Key words

- material
- properties
- soft
- flexible
- rigid
- rough
- smooth
- dull
- shiny
- transparent
- opaque
- absorbent
- waterproof
- fabric

Have a go! 📄

Sort these objects into hard and soft materials.

wool

pebble

wood

mirror

shell

feather

fabric

teddy

Science challenge

1 Match each material to the correct properties by drawing lines between them.

A glass	**1** bendable with force, strong
B foam rubber	**2** transparent, brittle
C metal	**3** soft, squashable

2 Choose the best material for the job! Join the material to the object.

A window	**1** metal
B door	**2** glass
C car	**3** paper
D packaging	**4** plastic
E dress	**5** fabric
F magazine	**6** wood
G doll	**7** expanded polystyrene

Scientific enquiry

Look around your home for items made from more than one material.

Fill in the table below, or construct a table with the headings shown on a separate piece of paper if you need more space.

Item	Materials	Useful property of material

Testing materials

Key facts

Scientists investigate the **properties** of **materials** by performing **fair tests** on them. They select a range of materials and then choose a property to test. They then complete the fair tests and use the results to compare that property in each material.

For example, Amandeep chose three materials – wood, plastic and rubber – for an investigation. All the samples of the materials were identical. They had the same length and thickness. The property Amandeep chose to test was flexibility. For this test she securely attached each material with a clamp to the end of a table. She set up the materials so that the same length hung over the table. She then hung a **weight** on each material at the same distance from the end. She recorded the result of each test as a picture like this one.

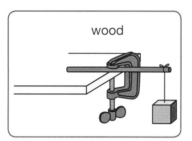

wood

When we view the three pictures together we can compare the property of flexibility in the materials.

Scientists use fair tests to investigate a wide range of properties, from the absorbency of different towels to the stretchiness of different makes of tights.

Have a go!

When Shazia and John tried the test in the **Key facts** box they produced a picture like the one shown above and also these two pictures below.

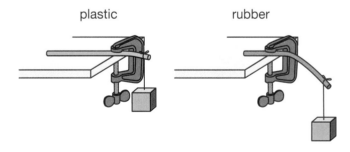

plastic   rubber

a Their teacher said that the test was not fair. Explain why. ....

....

....

....

b What do you think Shazia and John concluded from their test?

Science challenge

Caspar and Ravi want to test ladies' tights to see which kind are the most stretchy. They decide to take tights of the same length and measure how long one leg is. They will then put a weight into that leg and measure the new length.

a Circle the thing they will change in the test.

the length of the tights at the start   the weight   the kind of tights

b What should they use to measure the length of the tights?

....

c Here is a table of their results.
Complete the bar chart to show the results.

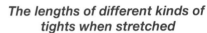

The lengths of different kinds of tights when stretched

Name of tights	Length of one leg when weight is added in cm
Comfy	120
Feelgood	110
Value	100
Wow	130

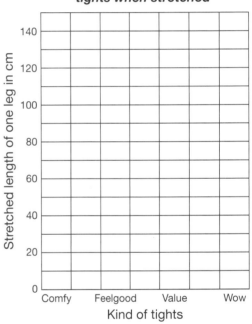

d Which tights were the most stretchy?

....

Scientific enquiry

If a material is waterproof it will not let water pass through it. This means that if you placed a card under the material and poured water onto the material, none of the water would pass through and make the card wet. Use this information to plan a fair test to compare the waterproof property of wool, cotton and plastic.

Rocks

Key facts

There are many different kinds of rock and each one has a set of **properties** just like other **materials**. Rocks are made of **particles** that have different colours, which give the rocks their various colours. Some rocks have particles that are easy to see, but for other rocks you need a magnifying glass to see their particles. Some rocks are very hard, such as granite, basalt and flint. They cannot be scratched easily by other rocks. Some rocks are easy to scratch, such as sandstone and limestone. Other types are crumbly, such as chalk. Some rocks will not let water into them but others, such as chalk, will let water in – they are called **permeable** rocks. These rocks have gaps between their particles, which is why they let the water in. Limestone and sandstone are also permeable rocks but flint, granite and basalt are not.

The properties of a rock make it suitable for different uses. We use granite for steps and buildings because it is very hard and does not wear away easily. Sandstone is not as hard and strong as granite but it has an attractive appearance and is very useful for building walls. Marble is a hard rock with attractive patterns in it and we use it to make floors and sculptures. Slate is a rock that can be split into thin sheets that do not let water pass through them. It is used to make roof tiles in some countries.

    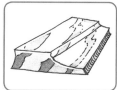

| chalk | slate | granite | sandstone | marble |

Key words

- properties
- particles
- flint
- granite
- basalt
- limestone
- permeable
- sandstone
- slate
- marble
- chalk
- crumbly

Have a go!

Match the rocks to their properties by drawing lines between them.

A granite	**1** hard wearing and has attractive patterns
B slate	**2** does not wear away
C marble	**3** can be split into sheets
D sandstone	**4** strong and attractive

Science challenge

For each rock, tick the box if you agree with the description.

Rock	Very hard	Can be scratched with another rock	Permeable	Crumbly
flint				
sandstone				
granite				
chalk				
basalt				
limestone				

Scientific enquiry

Look at the two rocks shown under magnifying glasses.

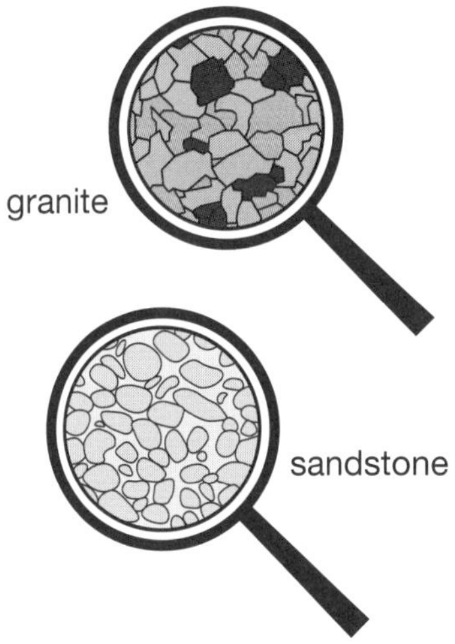

a Which rock is permeable? Explain your answer.

....

....

granite

sandstone

b Which rock is the stronger? Explain your answer.

....

c Find two rocks used in your home. Draw their appearance and describe their properties. Can you name these types of rock?

Rock A

Rock B

Solids, liquids and gases

Key facts

Solids, **liquids** and **gases** are three different forms of **matter** and they have different **properties**. Solids have a definite shape and **volume** and they do not flow. Volume means the space a certain amount of matter takes up. Liquids flow and take on the shape of the container so they do not have a definite shape but they have a definite volume. Gases flow even more easily than liquids and can flow in all directions. They spread out to fill the whole of the container into which you put them, not just the lower part that liquids fill. Gases can also be squashed. For example, when you use a pump to blow up a bicycle tyre the pump takes in air and squashes it into the tyre. As more and more gas is pumped in, the squashed air pushes outwards and makes the tyre hard.

Solids and liquids are easier to recognise because we can see them but most gases are colourless and we can not see them. We think of air as a gas but it is actually a mixture of gases. It contains oxygen, which we need to keep us alive, carbon dioxide, which we use to make drinks fizzy, and nitrogen, which we put into food packets to stop food going stale. Two other gases we might meet in our everyday lives are helium, which is lighter than air and we use it to make party balloons float, and **natural gas**, which we use in many cookers and can see as it burns to provide heat.

Have a go!
Solid, liquid or gas? You choose!

Write the names of the materials in the correct shape.

air	carbon dioxide	cheese	helium	metal
milk	orange juice	oxygen	plasticine	
pottery	syrup	water	wood	

solid

liquid

gas

Key words

- solid
- liquid
- gas
- matter
- volume
- stale
- flow
- air
- oxygen
- carbon dioxide
- nitrogen
- helium
- natural gas

Science challenge

diesel	liquid	air	ice
gas	milk	solid	

1 Use one of the words from the box to complete each of the sentences.

a A piece of granite is an example of a ....

b Blood is a red ....

c The mixture of gases around you is called ....

d .... is a solid until it melts.

e A liquid fuel trucks and some cars use is called ....

f The oxygen that divers use underwater is a ....

2 What is the gas we use in:

A? ....

B? ....

C? ....

D? ....

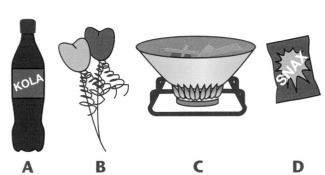

A   **B**   **C**   **D**

Scientific enquiry

We measure the volume of something in cubic centimetres, cm³. We measure the volumes of liquids in a measuring cylinder. The height that the liquid reaches on the scale shows you the volume.

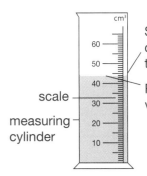

Sometimes the water curls up the edges of the cylinder. Ignore this bit.

Read off the water volume here.

scale

measuring cylinder

What are the volumes of the liquids in these measuring cylinders?

A ....

B ....

C ....

Heating and cooling

Key facts

Solid, liquid and **gas** are the three states of **matter**.

Heating and cooling can change states of matter from one to another. For example, if we heat a solid it changes into a liquid in a process called melting. The **temperature** at which a solid melts is called its melting point. If a liquid cools down it changes into a solid in a process called **freezing**. The temperature at which a liquid freezes is called its freezing point. A liquid's freezing point is the same temperature as the melting point of the solid that it forms. The melting point of ice is 0 °C and the freezing point of water is 0 °C.

If we heat a liquid strongly it starts to bubble as it changes into a gas. This process is called **boiling**. The temperature at which a liquid boils is called its boiling point. The boiling point of water is 100 °C. If a gas cools down it changes into a liquid in a process called **condensation**.

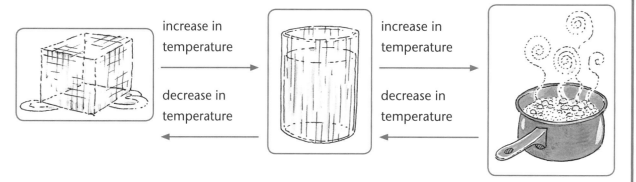

increase in temperature

decrease in temperature

increase in temperature

decrease in temperature

The pictures show what happens as water heats and cools. All these changes are **reversible changes**. When water changes into a gas it forms steam. This is a colourless gas but almost as soon as it rises into the air it cools and condenses to form a cloud of tiny water droplets.

Key words

- solid
- liquid
- gas
- matter
- temperature
- melting
- boiling
- freezing
- condensation
- reversible change

Have a go!

a What happens if you cool a liquid down until it is below its freezing point?

....

b Describe how you can get the liquid back again. ....

Science challenge

1 Sebastian leaves some ice cubes in a room at 20 °C.

   a What will the ice look like after 24 hours? ....

   b Explain its appearance. ....

   c What temperature will it be? Circle one.

         | 0 °C   10 °C   20 °C   24 °C   100 °C |

   d By how much must Sebastian raise the temperature to make steam? ....

2 a The freezing point of some lava (melted rock) in a volcano is 1000 °C.

      What temperature is the melting point of the rock? ....

   b The melting point of iron is 1535 °C. What is its freezing point? ....

   c Is the boiling point of iron above or below 1535 °C? ....

   d If some volcano rock and iron are at 1100 °C would you expect them

      both to be in the same state? Explain your answer. ....

      ....

      ....

Scientific enquiry 📄

Ice cubes remain solid in a freezer, but melt in a fridge and in the kitchen.
How long does it take? Devise a fair test to find out. In your plan write down:

a the materials you would use

b the factors you would control

c your prediction

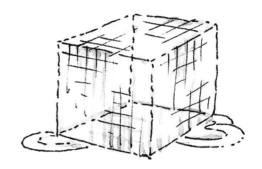

d an explanation of your prediction

e the measurements you would make.

Evaporation and condensation

Key facts

Gases form when **liquids evaporate**. All liquids can evaporate and they do this over a wide range of **temperatures**. Even a saucer of water left in a room at 20°C will slowly evaporate, but the higher the temperature the faster evaporation will be.

When you use a hair dryer the warm temperature of the air speeds up evaporation. The speed of the air also affects evaporation. The faster the air moves the more quickly evaporation takes place. When you hang washing on a line it is the warmth of the Sun that makes the clothes dry quickly.

When we heat water to 100°C it is evaporating so fast that bubbles of gas form in the water and we say it is **boiling**. The **water vapour** escaping at this high temperature is called steam. It is colourless and we cannot see it – just like water vapour at lower temperatures – but when it rises into the air it cools and **condenses**. This means it turns back into liquid water in the form of tiny droplets. Water vapour at lower temperatures can also condense if it touches a cold surface such as a window pane in winter.

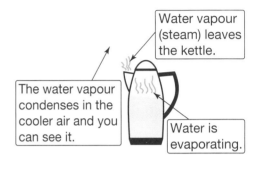

Water vapour (steam) leaves the kettle.

The water vapour condenses in the cooler air and you can see it.

Water is evaporating.

Key words

- evaporate
- water vapour
- boiling
- condense

Have a go!

1 Which of these statements are true and which are false? Write 'T' or 'F' in each box.

   a Water vapour is a gas. ☐   **b** Water evaporates above 100°C. ☐

   c If air cools down then evaporation slows down. ☐   **d** Condensation occurs when a liquid turns into gas. ☐

   e More liquid will condense on a cool surface than a warm one. ☐

2 a Which of these hair dryer settings will dry hair fastest? Tick (✓) one.

   low heat, slow fan ☐   high heat, slow fan ☐

   low heat, fast fan ☐   high heat, fast fan ☐

   b Explain why you made this choice. 📄

Science challenge

What will happen to this salty water solution? Explain on the lines below.

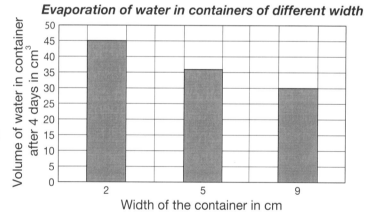

salty water

beaker

salty water

....

....

....

....

....

Scientific enquiry

Different groups in a class investigated different factors that could affect how quickly water evaporates. The bar charts show the results.

1 Answer these questions using the bar chart to the right.

a Which container had the least water in it?

b How much more water was in the container with the most water?

c What can you conclude from the bar chart?

d Explain your conclusion.

Evaporation of water in containers of different width

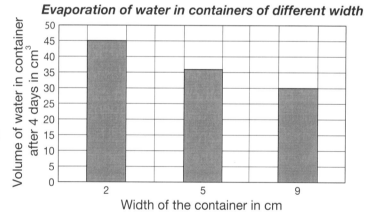

2 Answer these questions using the bar chart to the right.

a What can you conclude from the bar chart?

b What do you predict would happen if the class repeated the experiment with the fan running fast at a high heat setting?

Evaporation of water with different wind speeds

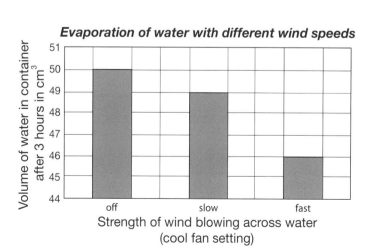

Explaining changes of state

Key facts

Water is vital to life and in most places is a feature of the weather. It can change state frequently and as it does so it moves between the ground and the air. The pattern of changes is known as the water cycle.

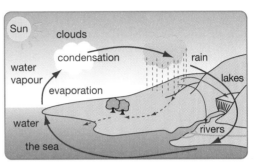

When air rises it cools and the **water vapour** in it **condenses** to form tiny water droplets. These make clouds. At the top of some clouds it is so cold that the droplets **freeze** and become ice crystals. In certain cold conditions they may fall as snow flakes. Usually the ice crystals melt again and form droplets, which in time join together to form rain drops and fall to the ground. The rain drops may collect into puddles. **Evaporation** takes place at their surfaces as **liquid** water changes into water vapour and mixes with other **gases** in the air.

All **matter** is made from tiny **particles** called **atoms** and **molecules**. In **solids**, such as ice, the particles hold onto each other firmly and cannot move about. In liquids, such as water, the particles do not hold onto each other and can slide over each other. In gases, such as water vapour, the particles can move away from each other and spread out in the space around them.

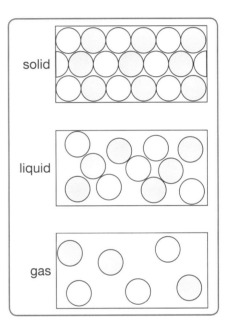

Key words

- water cycle
- atom
- molecule
- water vapour
- condense
- solid
- liquid
- gas
- particles
- evaporate

Have a go!

Which of these sentences are true and which are false ? Write 'T' or 'F' in each box.

a A liquid can evaporate to form a gas. ☐

b A liquid can condense to form a solid. ☐

c A solid can freeze to form a gas. ☐

d A gas can condense to form a liquid. ☐

e A liquid forms when a solid melts. ☐

f When a liquid freezes a solid forms. ☐

Science challenge

1 Use these words to complete the
sentences below.

rivers	cycle	atmosphere	droplets
down	drops	evaporates	surface
vapour	wind	condenses	water	flows

Water at the .... of the sea ....

and changes to water .... This rises in the air and

cools .... The cool ....

vapour .... to form water ....

Their huge numbers form clouds, which the ....

moves across the sky. When the clouds come over land the droplets may

join together to form rain .... that fall to the ground.

The water gathers to form streams and .... and

.... through the land back to the sea. This movement

of water between the ground and the .... is called the

water ....

2 Circle the correct word or phrase in each sentence.

a The particles in a solid / liquid / gas are held firmly together.

b When a solid melts, its particles stick together / slide over each other /
move apart.

c When a gas condenses / evaporates / boils to form a liquid, its particles
move apart / slide over each other / stick together.

d When a liquid evaporates / freezes / condenses to form a gas, its
particles stick together / slide over each other / move apart.

Scientific enquiry 📖

a A teacher places an upturned glass over a dish of water in
a sunny window. After an hour she sees water droplets on
the inside of the glass. Explain how they got there.

b If she set up a second dish of water and glass in
a shady place, what would happen?

Mixing and dissolving

Key facts

A mixture is when two or more **materials** are jumbled together. Sand and rice together make a mixture of two **solids**. Flour and raisins are another example of a mixture of two solids.

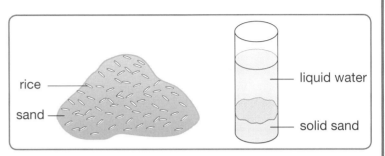

When you mix a solid with a **liquid**, different things can happen. The solid might sink to the bottom of the liquid, or it might make the liquid go cloudy. Some solids seem to disappear when they are mixed with liquids. For example, when salt is mixed with water it seems to disappear, but you know it is still there because the water tastes salty. The salt has **dissolved** – it has split up into such tiny **particles** that you can no longer see them. A mixture of a solid that has dissolved in a liquid (such as water and salt) is called a **solution**. If the solid has a colour, the solution will also have that colour, but you will always be able to see through a solution. The **substance** that dissolves to make the solution is called the **solute** and the substance in which the solute dissolves is called the **solvent**.

Key words

• dissolved • solution • solid • liquid • particles • solute • solvent

Have a go!

1 What is a mixture? Tick (✓) the best answer.

a A large pile of material ☐

b Sand and rice together ☐

c Two or more materials jumbled together ☐

d When a solid is added to a liquid ☐

2 Robert adds some paper clips to some water.

a What is the solid in the mixture? ....

b What is the liquid in the mixture? ....

3 Circle the solids that will dissolve in water.

instant coffee granules   marbles   metal coin   rice   salt   sand   sugar

Science challenge

Look at the line graph, which shows how stirring speed changed the time sugar took to dissolve.

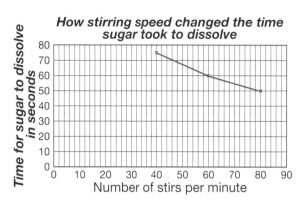

How stirring speed changed the time sugar took to dissolve

a What factor was the scientist investigating?

....

b Describe what the graph tells you. ....

c At 90 stirs per minute, the sugar dissolved in 45 seconds. Mark this on the graph and join your point to the others with a straight line.

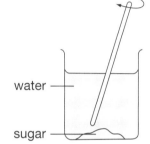

Scientific enquiry

A class did four investigations to find out how different factors affect how quickly sugar dissolves. The results are shown in the tables.

Stirring speed	Time taken to dissolve in seconds
slow	70
medium	50
fast	34

Sugar type	Time taken to dissolve in seconds
powder	35
granules	45
cubes	70

Temperature in °C	Time taken to dissolve in seconds
20	45
30	25
40	20

Volume of water in cm³	Time taken to dissolve in seconds
20	45
40	40
60	35

a One factor that the class investigated was the type of sugar. Write down the four things that the learners needed to keep the same, to do a fair test.

....

....

b Complete this sentence to write the conclusion for this investigation:

The finer the sugar was, the more .... the sugar dissolved.

c Write down the conclusions for the other investigations.

Bar charts and line graphs

Key facts

Bar charts and line graphs help scientists to see clearly how changing one factor in an investigation affects another **factor**. Bar charts and line graphs both have a horizontal line running along the bottom, called the *x*-axis, and a vertical line running up the left-hand side, called the *y*-axis. The *x*-axis shows the factor that you are changing in the investigation and the *y*-axis shows its effect on the other factor.

In a bar chart rectangular bars rise vertically from the *x*-axis. You use a bar chart when the factor you are investigating does not fit on a scale. For example, you might want to find out how many different pets eat meat, eat plants, or eat plants and meat. Above is a bar chart showing the results of a class survey about pets.

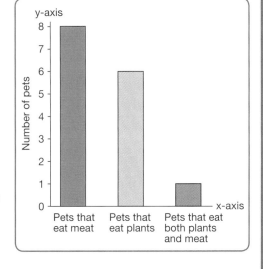

In a line graph there is usually only one line in the space between the *y*-axis and the *x*-axis. The line may be curved or straight or a mixture of the two. You use a line graph when the factor you are investigating does fit on a scale. Here is a line graph showing the results of an investigation.

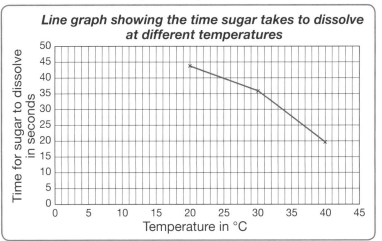

Line graph showing the time sugar takes to dissolve at different temperatures

Have a go!

1 Answer these questions about the bar chart above.

   How many pets eat:
   a meat   **b** plants   **c** both plants and meat?

2 Answer these questions about the line graph above.

   a How many seconds did it take the sugar to dissolve at 25 °C?

   b What was the temperature at which the sugar dissolved in 25 seconds?

Key words

- bar chart
- line graph
- factor

Science challenge

To be sure the results of an investigation are correct it is always best to repeat your measurements. Here is a line graph showing the results of an investigation. When the scientist repeated it, one result she recorded and plotted on the graph did not fit with the line. When this happens, the scientist has to think of an explanation for the result that did not fit in with the others and check the procedure of the investigation carefully and look for an error.

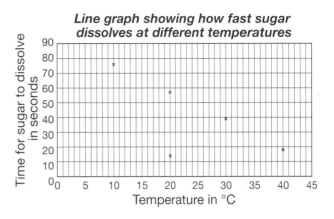

Line graph showing how fast sugar dissolves at different temperatures

a Which of the following could explain the point that does not fit on the line? Tick (✓) two boxes.

i The stopwatch stopped by mistake and no one noticed. ☐

ii A much smaller amount of sugar was used than should have been used. ☐

iii The person doing the time forgot to stop the stopwatch. ☐

iv A much bigger amount of sugar was used than should have been used. ☐

b Why is it best to repeat measurements when you do an investigation?

....

c At which of these temperatures do you predict that salt will dissolve quickest in water? Circle one.

0 °C   10 °C   20 °C   40 °C   60 °C

d Why do you think this? ....

Scientific enquiry

Amita set up an investigation to test the prediction that three different types of bean plants will grow to the same height ten days from germination. Here are the results.

Type A 15 cm   **Type B** 20 cm   **Type C** 10 cm

a Make a bar chart of these results.

b Do the results agree with the prediction? ....

Explain your answer. ....

....

Separating mixtures

Key facts

You can separate some **solids** from each other by using a sieve. It works when one part of the mixture is made of larger pieces than the other – for example, to separate a mixture of raisins and flour.

You can also separate a **liquid** from tiny pieces of solid if you can still see the solid in the liquid – for example in a mixture of sand and water. When you pour the mixture through a **filter**, the liquid passes through into another container, but the solid stays in the filter paper. This is called filtration. The **substance** that is left in the filter paper is called the residue. The liquid that has passed through the filter paper is called the **filtrate**.

You can use **evaporation** to get a solid back from a **solution** in which it is **dissolved**. Simply leave the solution in a dish in a warm place.

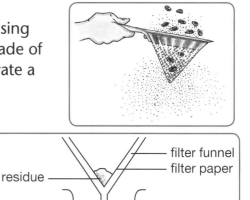

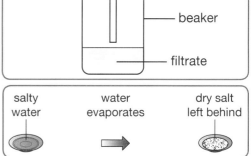

residue — filter funnel
filter paper

beaker

filtrate

salty water → water evaporates → dry salt left behind

Key words

- sieve
- filter
- solid
- liquid
- filtrate
- residue
- filter funnel
- filter paper
- beaker
- evaporation
- solution
- filtration

Have a go!

1 Which of these mixtures could you separate using filtration? Mark them with a tick (✓).

a salt and water ☐

b sand and water ☐

c leaves in water ☐

d pebbles in water ☐

e instant coffee granules in water ☐

f paper clips and water ☐

g sugar and water ☐

h chocolate chips and sugar ☐

2 Which of the mixtures in question 1 could you separate by sieving?

....

Science challenge

a The words in the box show the equipment you might use to separate a mixture of gravel and water. Draw a flow chart in the box below to show how you would do this.

> filter paper   funnel   empty beaker
>
> beaker containing mixture of gravel and water

b Explain in words how the gravel and water are separated. ....

....

Scientific enquiry

The diagram shows a solution in a dish being heated. Read the labels carefully then answer the following questions.

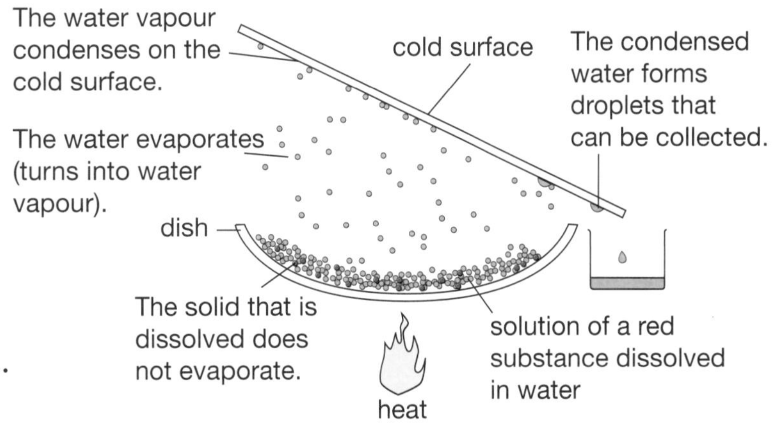

The water vapour condenses on the cold surface.

cold surface

The condensed water forms droplets that can be collected.

The water evaporates (turns into water vapour).

dish

The solid that is dissolved does not evaporate.

heat

solution of a red substance dissolved in water

a What is in the solution being heated?

....

b What is the likely colour of the solution? ....

c Which substance in the solution evaporates? ....

d What will be left in the dish when all the liquid has evaporated?

....

e Circle the reversible changes that are used to separate the parts of this mixture.

> freezing   condensation   boiling   melting   evaporation   dissolving

Reversible and irreversible changes

Key facts

Some changes, such as melting and **freezing**, are **reversible**. This means that the **material** can change back to its original form after the change.

Many changes are **irreversible**. This means that the material cannot change back into its original form. Cooking, burning and chemical reactions are examples of irreversible changes:

- If we cook a raw egg, it changes from a runny liquid to a solid.

- When melted wax in the candle wick burns, it takes part in a change with oxygen in the air and produces carbon dioxide and water vapour – two colourless gases.

- When we add bicarbonate of soda to vinegar, bubbles of carbon dioxide are produced. When we add water to cement or to plaster of Paris, a hard solid is produced.

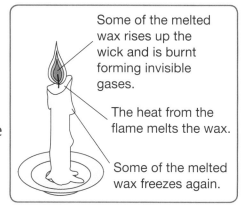

Some of the melted wax rises up the wick and is burnt forming invisible gases.

The heat from the flame melts the wax.

Some of the melted wax freezes again.

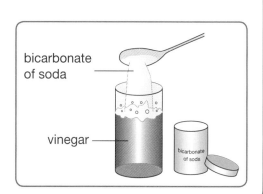

bicarbonate of soda

vinegar

Key words

- reversible   • irreversible   • wick   • reaction   • solid   • liquid   • gas

Have a go!

1 Circle the irreversible changes:

baking a cake	boiling an egg yolk	making plaster of Paris
adding sugar to water	freezing an egg yolk	melting iron

2 Look at the drawing of the candle above.

a What two reversible changes happen in a candle?

b What irreversible change happens in the candle?

3 After you had added bicarbonate of soda to vinegar, explain how you would know when the change had finished.

Science challenge

Fill in the gaps using the words below.

solid   liquid   gas   a reversible   an irreversible

Shazia is busy in the kitchen where she is making a cake.

a She melts butter in a pan, turning it from a .... to a

.... If she cooled the butter again, it would turn back

into a .... That means this is ....

change.

b She decides to have a cup of tea and boils a kettle. Some of the water

changes from a .... to a ....

as the kettle boils, filling the kitchen with steam. As the steam hits the

window, it condenses and turns back from a .... to a

.... That means this is .... change.

c She decides to have poached eggs for her lunch. As she cracks them into

the pan, they are a .... Once they are cooked, they

are a .... This is .... change.

Scientific enquiry

Ahmed measured the mass of a burning candle every minute for ten minutes and plotted the results on a graph.

a Join up the points on the graph.

b What does the graph show?

c Does the graph show a reversible or irreversible change?

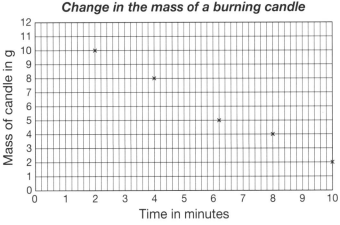

Change in the mass of a burning candle

d Identify a point that does not match the pattern shown by the other points by drawing a circle around it.

e Suggest how Ahmed can make the investigation more accurate.

1   Ahmed is making a survey of the properties of five materials.

Here is a table of his results.

Material	Soft / hard	Smooth / rough	Dull / shiny	Flexible / rigid	Transparent / opaque	Absorbent / waterproof
A	soft	smooth	dull	flexible	opaque	absorbent
B	hard	rough	dull	rigid	opaque	waterproof
C	hard	smooth	shiny	rigid	transparent	waterproof

(a) Which two materials have the most differences? .... [1]

(b) What could be the identity of each material? Match the letters to the material names by drawing lines between them.

A		glass
B		cloth
C		granite

[3]

2   Joanne is investigating the waterproof properties of three pieces of material.

Here is a table for her predictions.

Material	Prediction
plastic sheet	
cotton shirt	
wool coat	

(a) Fill in the prediction column. [3]

(b) She plans to pour water onto each material and measure the diameter of the water spot made underneath it. Describe how she can make her test fair.

....

.... [2]

3 **(a)** Name three gases. .... [3]

(b) In the box draw the arrangement of particles in a solid.

[1]

(c) In the box draw the arrangement of particles in a gas.

[1]

4   Match the state of matter to its description by drawing lines between them.

State of matter

solid

liquid

gas

Description

flows but has a definite volume

flows but can change its volume

has a definite shape and volume

[3]

5   Shazia set up three containers of water at different temperatures. She placed 50 cm³ of water in each container and left them for ten minutes before she measured the volumes again.

Here is a bar chart of her results.

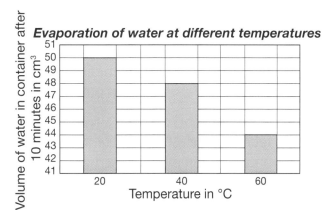

Evaporation of water at different temperatures

Volume of water in container after 10 minutes in cm³

Temperature in °C

(a) How much water did each container lose? Fill in the table.

Container	Volume lost in cm³
containing water at 20 °C	
containing water at 40 °C	
containing water at 60 °C	

[3]

She left each container open to the air and none of them tipped over.

(b) By what process did the water leave? .... [1]

(c) What did Shazia conclude from her investigation? ....

.... [1]

6   Tick (✓) to show if a change is reversible or irreversible.

Process	Reversible	Irreversible
burning candle		
boiling water		
boiling an egg		
freezing water		

[4]

Physics

Forces

A **force** is a push or a pull. Here are some examples of pushes and pulls and what they do.

You push a ball when you throw it or kick it. This makes the ball start moving. When you catch a ball your hands push on the ball. This makes the ball stop moving. If you kick a moving ball the push can make it change direction. It can also make the ball move faster. When you are speeding along on a bicycle you can press on the brakes so they push on the wheels and slow you down. Even when you are using modelling clay you use forces – pushes and pulls to make the clay into the shape you want.

Some objects such as an elastic band or a metal spring stretch when they are pulled. They build up a pulling force too, which you can feel. For example, if you stretch an elastic band between two fingers you can feel it dig into your skin as you keep on stretching it. When you let go, the elastic band goes back to its original shape.

An instrument called a forcemeter uses the stretchiness of a metal spring to measure forces. The unit that we measure a force in is the newton. The indicator on the forcemeter moves along the scale to show you the size of the force.

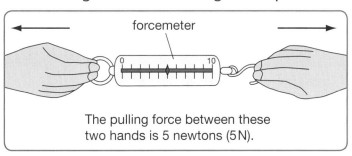

The pulling force between these two hands is 5 newtons (5N).

Key words

- elastic
- spring
- forcemeter
- newton
- push
- pull

Have a go!

1 What can forces do? Circle the things that forces can do.

make things move	make things change colour	make things slow down
make things change direction	make things go cold	make things speed up
make things go dim	make things change shape	

2 What units do we measure forces in? Circle the correct one.

neutrons	newtons	newtowns	pushes and pulls

Science challenge

1 We use an arrow to show the direction of a force. This donkey is pulling a cart. The arrow shows the direction of the pulling force.

  a Draw in the arrow showing the direction in which this force is acting when the girl kicks the ball.

  b Is the force a push or a pull? ....

2 This girl is exerting a force on the chair.

  a Draw an arrow to show the direction in which the force is acting.

  b Is the force a push or a pull? ....

Scientific enquiry

It is important to repeat measurements to make sure that they are correct, but measurements of the same thing can be slightly different. Claude investigated the stretching of an elastic band. He put the 0 of the ruler at the bottom of the elastic band and then attached a weight and measured the stretch. He repeated each measurement twice. The graph shows his results.

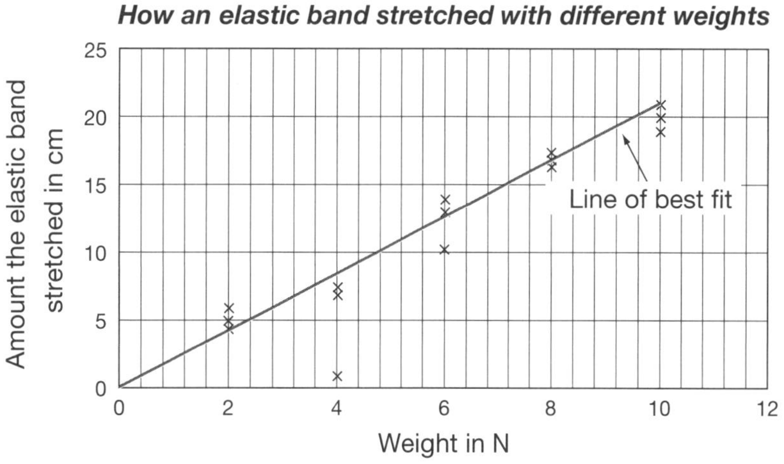

How an elastic band stretched with different weights

Amount the elastic band stretched in cm (y-axis, 0 to 25)
Weight in N (x-axis, 0 to 12)

Line of best fit

a What causes the elastic band to stretch? ....

b Which set of three measurements seem the most accurate? ....

c Explain your choice. ....

d What does the graph show? ....

....

Mass, weight, energy and movement

Key facts

The amount of **matter** in a **substance**, such as a block of wood or a glass of water, is called its **mass**. Scientists measure mass in units called grams and kilograms (1000 grams).

Gravity is a **force** between any two objects in the universe. We only notice its pull when one object (the Earth) is very much larger than the other (you). In this case the larger object pulls the smaller object towards its centre. The pulling force of gravity on the mass of an object is called **weight**. Scientists measure weight in newtons but in everyday life people weigh the mass of an object in grams and kilograms.

There are two kinds of **energy** – stored energy and moving energy. If you wind up a clockwork toy you store energy in its spring. When you let it go, the spring unwinds and the energy changes to movement energy and makes the toy move along. Cars and trucks use **fuel** as an energy store and when their engines burn it, it changes to moving energy. Energy is stored in food and you change some of it to moving energy when you use your **muscles**.

Key words

- mass
- gravity
- grams
- kilograms
- energy
- matter
- substance
- weight
- newtons

Have a go!

1 Draw two arrows to show how an apple would fall if it were dropped at the two places shown.

2 Use the following words to complete the sentences below.

| gravity | mass | Earth | matter | weight |

The amount of .... in an object is called

its .... The force of .... pulls

on the mass of an object. This pull gives the object

its .... which is also a force. We usually think of this force

pushing down on anything below it, such as the ....

Science challenge

1 What are the weights of these objects?

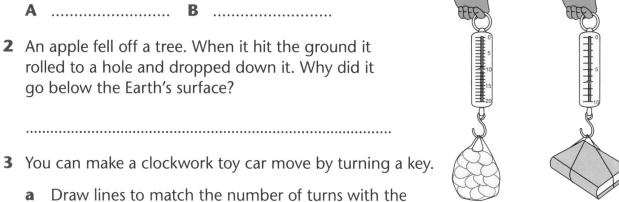

A .... B ....

2 An apple fell off a tree. When it hit the ground it rolled to a hole and dropped down it. Why did it go below the Earth's surface?

....

3 You can make a clockwork toy car move by turning a key.

a Draw lines to match the number of turns with the distance the car travels.

Number of turns	Distance travelled
4	50 cm
1	200 cm
3	100 cm
2	150 cm

b Describe how the energy in the toy changes when you let the toy car go.

....

c Do the wheels on a toy car push or pull on the floor to make the car

move forwards? ....

Scientific enquiry

1 A mass of one kilogram has a weight of about 10 newtons.

a What is the weight of a 1 kilogram bag of flour in newtons? ....

b What is the weight of a 400 g tin of peas in newtons? ....

2 Look at other items in a kitchen that show their weight in kilograms and grams, and convert them to the scientist's unit of weight – the newton. Write down four examples in the table.

Item	Mass in kg and g	Weight in newtons

Friction and air resistance

Key facts

Friction is a **force** that acts when two surfaces rub together.

When you walk along, the soles of your shoes touch the ground and push back on the ground to move you forwards. When this happens a friction force acts in the opposite direction. The strength of this force is equal to your pushing force and you move along. If the friction force was weaker you would slip. Rough surfaces make strong friction forces and smooth ones make the friction force weaker. When a floor is wet, water fills the tiny hollows in its surface and makes it smoother. This makes the friction force weaker and when you walk on the surface you can slip. When an object slides, friction acts to slow it down – otherwise it would keep moving forever!

force of push on the ground

friction force

Air resistance is a force that acts when an object moves through the air. The object pushes on the air as it moves and the air pushes back on it with a force called air resistance. An object such as a parachute, which has a big surface pushing on the air, generates a large amount of air resistance and this slows it down. Racing cars and planes are pointed at the front to cut through the air easily and generate as little air resistance as possible. This kind of shape to reduce air resistance is called a streamlined shape.

Key words

- friction
- air resistance
- force
- streamlined
- push
- pull

Have a go!

1 **a** Why do tractor tyres need rough surfaces to get them over the

   ground? ....

   b What would happen if the tyres had smooth surfaces? ....

   ....

2 Which will create the most friction? Number them 1 to 3, with 1 for the most friction.

an ice cube pushed over a mirror ☐   an eraser pushed over a thick carpet ☐

a toy car pushed over a smooth table ☐

Science challenge

Ahmed investigated how quickly spinners with different wing areas fell to the ground. The table shows his results.

Area of wings in cm²	Time to fall 2 metres in seconds
120	2
180	3
250	5
300	6

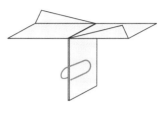

a Fill in the other three points on this graph.

b Draw lines to connect the points with straight lines.

c Describe the pattern that the graph shows.

....

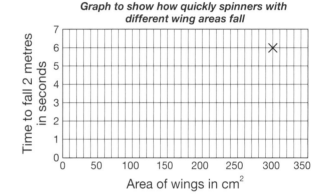

Graph to show how quickly spinners with different wing areas fall

d What two forces act on the falling spinner?

....

e Write an explanation for Ahmed's results. ....

....

....

Scientific enquiry

1 What do you predict would happen if you took two sheets of paper, squashed one up into a ball, then held both up in the air and dropped them together?

....

....

2 Carry out an accurate investigation to test your prediction, and describe it here.

....

....

3 What did you conclude? ....

....

Magnets

Key facts

A magnet has two ends called **poles**. When a magnet is free to move in a compass, one end points north and the other end points south. The ends of the magnet are known as the north and south poles.

If you put two magnets together so that their similar poles – two north or two south – come close you can feel the magnets pushing away from each other. We call this repulsion. If you put two magnets together so that they have two different poles coming close you will see them pull together. We call this attraction.

Magnets attract some metals such as iron, nickel and steel. This means that objects made of these **materials** are pulled towards a magnet and seem to 'stick' to it. Materials like these are called magnetic materials. Be careful – people think that all metals are attracted to magnets but this is not true. Gold, silver and copper, for example, are not attracted to a magnet and are called non-magnetic materials. A strong magnet can attract things made of magnetic materials through paper, card and fabric.

Key words

- magnet • poles • north • south • repulsion • pull • push • iron
- nickel • steel • magnetic materials • non-magnetic materials • attraction

Have a go!

Will the magnets be attracted to one another or will they repel one another?
Write 'A' for 'attract' or 'R' for 'repel' in each box.

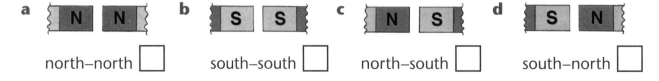

a N N   b S S   c N S   d S N

north–north ☐   south–south ☐   north–south ☐   south–north ☐

Science challenge

1 When you put two 'like' poles together (two that are the same), what happens?

....

2 True or false? Write 'T' or 'F' in each box.

a If you put 'like' poles together, they will repel each other. ☐

b If you put opposite poles together, they will repel each other. ☐

c If you put 'like' poles together, they will attract each other. ☐

d All non-metals are attracted to a magnet. ☐

e Metals containing iron are attracted to a magnet. ☐

f If you put two north poles together, they will attract each other. ☐

Scientific enquiry

Su Lin wanted to find out if bigger magnets are stronger than smaller magnets. She tested four magnets by seeing how many paper clips each one could hold. The pictures show the maximum number of paper clips that each magnet could hold.

a Fill in this table for the results. Include the top paper clip when you count up the paper clips.

b Write down one thing that Su Lin kept the same to make this test fair.

....

....

Magnet	Size	Number of paper clips held
A	small	
B	large	
C	medium	
D	small	

c What did Su Lin conclude from her investigation? ....

....

....

The spinning Earth

Key facts

The Earth spins on its **axis**, which is tilted at an angle of 23° from the vertical. It is the spinning of the Earth that makes it seem as if the Sun moves across the sky. The axis is like an invisible line running through the centre of the Earth from **pole** to pole. The Earth spins round once every 24 hours. It is daytime on the part of the Earth that faces the Sun, and night on the part of the Earth facing away from the Sun.

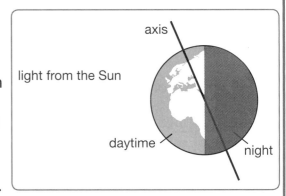

If you were to look down on the Earth from over the north pole you would see that it turns anticlockwise. Back on the surface this means that the Sun appears to move across the sky as the Earth turns. It rises on the eastern **horizon**, reaches its greatest height in the sky at midday, then sinks and sets on the western horizon. As the Sun's position in the sky changes, the shadows that objects cast change.

Key words

- axis
- north pole
- anticlockwise
- horizon

Have a go!

Connect the two halves of the sentences so that they make sense.

The Earth spins on its axis once	that causes night and day.
It is the Earth's rotation	it is daytime where you are.
When the side of the Earth where you are is turned towards the Sun	every 24 hours.
When the side of the Earth where you are is turned away from the Sun	it is night-time where you are.

Science challenge

We can use shadows to tell the time. We position sundials on raised stands or fix them to walls. The arm causes a shadow.

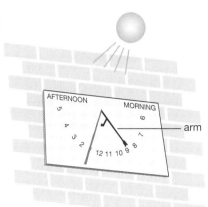

a What causes the shadow on the sundial?

....

b At what time of day will the shadow be the shortest? Tick (✓)one.

morning ☐   midday ☐   evening ☐

it does not change length ☐

c What is the time on the sundial above? ....

d On the picture below, draw in the shadow.

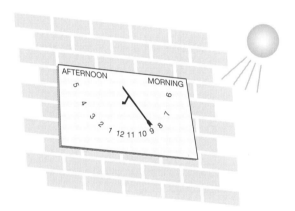

Scientific enquiry

Investigate how the direction of the shadow changes through the day. Try this with a friend.

a In the morning, mark your position on the ground and the position of your shadow.

b An hour later, mark the position of your shadow and predict where it will be in an hour's time. Mark the position.

c An hour later, mark the position of the shadow again and compare it with your prediction mark.

d Continue predicting and recording your shadow.

e Circle an answer. Do your predictions become more accurate / remain at the initial accuracy / become less accurate?

The Earth in its orbit

Key facts

The Earth moves round the Sun in an orbit. It spins round on its **axis** as it travels in its orbit. It takes the Earth one year to make a complete orbit of the Sun. At different places in the orbit, the Sun rises and sets at different times of day. The Earth keeps the tilt of its axis pointing in the same direction as it moves in its orbit. This helps to produce the seasons.

For example, when the axis in the northern hemisphere tilts towards the Sun there is more daylight and warmth reaching places there and it is summer, but when the axis tilts away there is much less daylight and warmth and it is winter there.

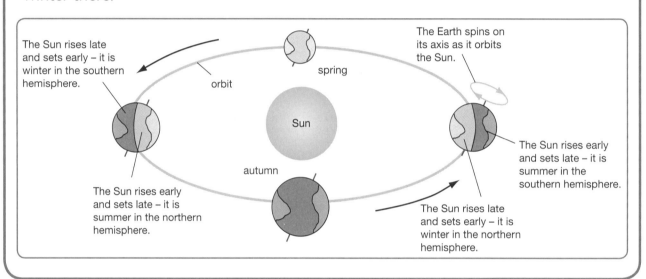

Have a go!

Choose the correct words from the box below to complete the sentences.

short	sets	long	dark	light
year	spins	bounces	orbits	late

Key words

- orbit
- tilt
- seasons
- hemisphere
- axis

a In the summer, the evenings are .... and light.

b In winter, it gets .... earlier.

c The Earth .... as it goes round the Sun.

d The Earth .... the Sun.

e It takes one .... for the Earth to orbit the Sun.

Science challenge

1 Scientists measure the time of sunrise and
sunset in all countries. They plot the times on
a graph. This graph show the times for sunrise
and sunset in Cardiff in Wales. This country,
like many in Europe (and others such as the
Middle East and Africa, parts of Australia, the
whole of New Zealand and North America),
uses daylight saving time, where the clocks are
put forward and back during the year. Answer
the questions about it below.

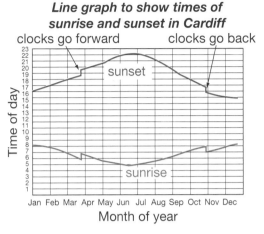

*Line graph to show times of
sunrise and sunset in Cardiff*

a What happens to the time of sunrise in the
first six months of the year? Circle your answer.

| it gets earlier | it stay the same | it gets later |

b What happens to the time of sunset in the first six
months of the year? Circle your answer.

| it gets earlier | it stay the same | it gets later |

c In which month does Cardiff gets most hours of sunlight? ....

d To the nearest hour, what is the highest number of daylight hours

Cardiff gets in a day? ....

e At what time is sunrise when Cardiff gets this number of hours of

daylight? ....

2 The Sun always rises in the east and sets in the opposite direction. Which

direction is this? ....

Scientific enquiry

Find out about the following scientists and match them to their discoveries.

Galileo	discovered galaxies beyond the Milky Way Galaxy.
Newton	was the first to suggest that the Earth was not the centre of the universe.
Copernicus	discovered the planet Uranus.
Hubble	discovered the moons of Jupiter.
Herschel	showed that the force of gravity held planets and moons in their orbits in the Solar System.

The way we see things

Key facts

Light comes from **light sources** such as the Sun, a torch or a candle. It travels in straight lines and cannot bend round things. We use our eyes to see things. We can see a light source when the light from it travels directly into our eyes. We show the way light travels into our eyes by drawing straight lines called light rays or light beams, as the first picture shows.

Sophie sees the light from the candle because it travels straight into her eyes.

light ray

Obviously we can also see things that are not light sources. We can do this because light bounces off objects and enters our eyes. This bouncing off objects is called reflection. In the second picture, the light hits the magazine, bounces off and enters the boy's eyes so he can see the magazine.

Scientists use mirrors to study reflection. When a light ray hits a mirror, it reflects off and changes direction.

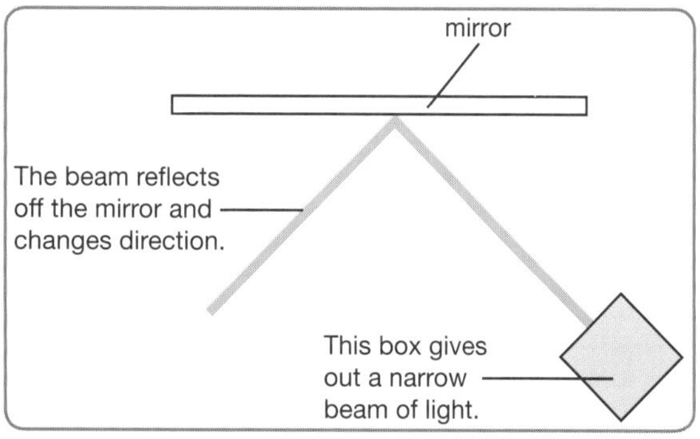

If you change the path of the beam, the path of the reflected beam also changes.

mirror

The beam reflects off the mirror and changes direction.

This box gives out a narrow beam of light.

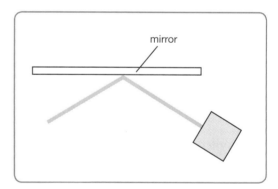

mirror

We describe the amount of light as its intensity. The light intensity is low at dawn and dusk but is high at midday in a cloudless sky. There are instruments that can measure the light intensity.

Key words

- light source
- light ray
- light beam
- reflection
- mirror
- sight
- eyes

Have a go!

On the two pictures of the mirror and light box above draw arrows to show the directions in which the light beams travel.

Science challenge

1 Look at this picture.

Draw direction arrows to show how the girl can see the spider.

2 Reuben and Nina investigated which materials they could use as mirrors. They found out which materials could reflect the light beam from a torch, and also which materials they could see themselves in. The table shows their results.

Material	Could it reflect a torch beam?	Could we see our reflection in it?
cardboard	yes	no
kitchen foil	yes	yes
silver spoon	yes	yes
woollen hat	yes	no

a What light source did they use in their investigation?

....

kitchen foil

b Which surfaces reflect light better – shiny or dull?

....

c Which materials from Reuben and Nina's investigation might we use to

make a mirror? ....

Scientific enquiry

a What do you predict you might see when you look in still water and shine a torch beam on it? What do you think you might see when you make waves in the water with a spoon then look in the wavy water and shine a torch beam on it? Record your predictions in the table.

Material	Could it reflect a torch beam?	Could you see your reflection in it?
still water in a cup		
wavy water surface made by spoon		

b If you predict a difference, explain it. ....

....

Shadows

Key facts

When light shines on an object, it may pass through it if the object is **transparent** like window glass. If the object does not let light through it, like a piece of wood or even you, we describe the object as **opaque**. When light strikes an opaque object the object stops the light rays from reaching the other side. Because other light rays passing either side of the object cannot bend around it, a dark area forms there because the light cannot reach there. This dark area is called a shadow.

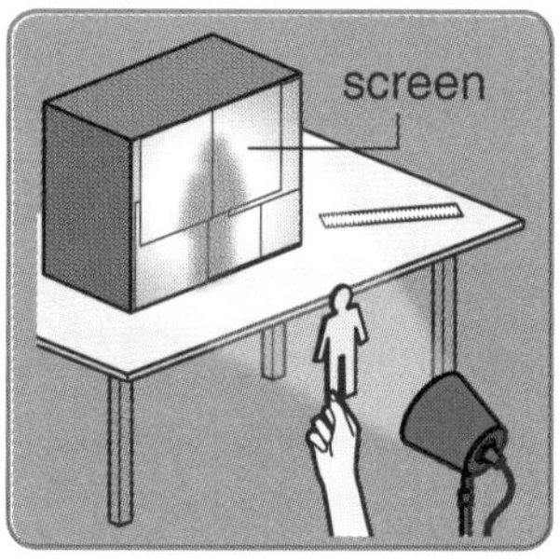

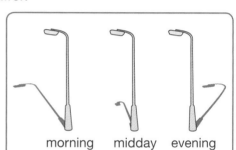

We can investigate shadows by using a **light source**, an opaque object such as a puppet and a screen as shown in the picture. The shadow has the shape of the object that is making it but when we move the object towards the screen and away from the light source, its shadow gets smaller. When we move the object away from the screen towards the lamp, its shadow gets larger.

Objects lit by sunlight cast shadows. As the Sun moves across the sky the direction of the shadow and its length change. The picture shows how the shadow of a street lamp changes during a sunny day.

morning   midday   evening

Key words

- transparent
- light source
- opaque
- sunlight
- shadow
- light ray
- screen

Have a go!

Sort the words in this word wall into the 'Transparent' and 'Opaque' columns.

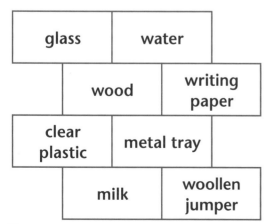

glass | water | wood | writing paper | clear plastic | metal tray | milk | woollen jumper

Transparent	Opaque

Science challenge

1 Look at the puppet in the picture opposite. What do you predict will happen to its shadow if we move it:

a nearer the screen? ....

b nearer the lamp? ....

2 Look at the shadows in the two pictures and in each picture circle the Sun that is making the shadow.

3 True or false? Write 'T' or 'F' in each box.

a Light travels in straight lines. ☐

b Light travels in wavy lines and can bend round things. ☐

c Opaque objects block light. ☐

d Transparent objects block light. ☐

e A shadow forms on the same side of the object as the light shining on it. ☐

Scientific enquiry

Describe how you and a friend could use a piece of chalk and a hard playground surface to find out how your shadow changed during the day.

....

....

....

Sound

Key facts

If you flick a ruler, it makes a sound as it wobbles up and down. This movement is called **vibrating**. When the ruler stops vibrating it stops making a sound.

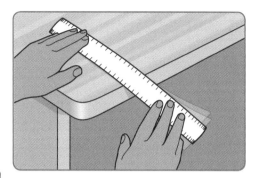

When something vibrates it sets the **material** next to it vibrating. This vibration forms a wave – a **sound wave** – that passes through the material. When sound waves move through the air they can pass into your ears, which detect them and send messages along **nerves** to your brain so that you can hear.

The loudness of the sound is due to how much the vibrating object moves up and down. If you beat a drum hard its skin vibrates greatly up and down and makes a loud sound. If you tap the drum it does not move up and down as much and makes a quieter sound.

If an object vibrates very quickly it makes a high sound or a high-pitched sound like the word 'ding'. If an object vibrates more slowly it makes a low sound or a low-pitched sound like the word 'dong'.

As sound waves travel through the air they lose their **energy** so if you stand a distance away from the sound source you might not hear them.

Have a go!

1 Write a number in each box to put these stages in hearing a sound in the correct order.

Key words

- vibration
- energy
- nerves
- sound wave
- pitch

a A sound wave enters the ear. ☐

b The brain receives a message and allows you to hear it. ☐

c A ruler vibrates on a table. ☐

d A nerve carries the message from your ear to your brain. ☐

e A sound wave travels through the air. ☐

2 Which two statements below are true? Tick (✓) two boxes.

a Small vibrations make quiet sounds. ☐   b Large vibrations make quiet sounds. ☐

c Large vibrations make loud sounds. ☐   d The size of a vibration does not affect the loudness of the sound. ☐

Science challenge

1 Describe the sound the following sound sources make. For each one, circle two words.

    a   Source makes large slow vibrations.

        | quiet   loud   low pitch   high pitch |

    b   Source makes small fast vibrations.

        | quiet   loud   low pitch   high pitch |

    c   Source makes large fast vibrations.

        | quiet   loud   low pitch   high pitch |

    d   Source makes small slow vibrations.

        | quiet   loud   low pitch   high pitch |

2 Look at the picture below. Where would the sound from the radio be loudest? Circle A, B or C.

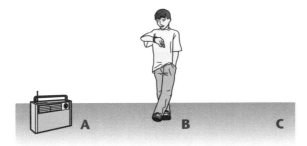

    Give a reason for your answer. ....

....

Scientific enquiry

Make a note of sounds you hear during the day and select sound sources to complete this table.

	Sound source	Quiet	Loud	Low pitch	High pitch
a		✓		✓	
b			✓	✓	
c		✓			✓
d			✓		✓

Sound and materials

Key facts

Sound waves pass through **solids**, **liquids** and **gases**. You can listen to a sound through a solid by simply putting the sound source on a table top and pressing your ear to the surface a short distance away.

You can listen to sound through a liquid by placing the sound source in a **waterproof** bag, as the picture shows, and listening to the side of the water-filled tank. The sound is loudest through solids, less loud through liquids and least loud through gases like air.

The loudness of sounds can vary greatly. Scientists use a sound level meter and a **decibel** scale to measure them.

Very loud sounds can damage the ears, so people who use or work near noisy machinery wear ear protectors. The ear protectors contain **materials** that absorb the **energy** in sound waves and stop loud sounds reaching the ears. Illnesses and infections can also affect hearing, but doctors can usually treat these problems.

Sound	Loudness in decibels
jet aircraft taking off	140
jet plane overhead	100
vacuum cleaner	70
normal speech	60
whisper	20
limit of normal hearing	0

Have a go!

1 If you listened to a sound through air, then water, then wood, which sound would be loudest and which would be the quietest?

Key words

- decibel
- sound waves

   a loudest ....   **b** quietest ....

2 **a** What does the picture show? ....

   b Look at the decibel scale and estimate the loudness in decibels.

   ....

3 Why do people using machinery wear special ear protectors?

   ....

Science challenge

True or false? Write 'T' or 'F' in each box.

a You should never poke anything in your ear, in case you cause damage. ☐

b The further away you are from a noise, the louder it is. ☐

c People wear ear protectors to make noises sound louder. ☐

d Loud noises can damage ears. ☐

Scientific enquiry

Adya and Dembe investigated which materials are best at soundproofing. They took a buzzer, covered it with a single layer of material and put it in a box. They then went into the school hall and found out how far away the box had to be before they could no longer hear the buzzer.

The bar chart shows their results.

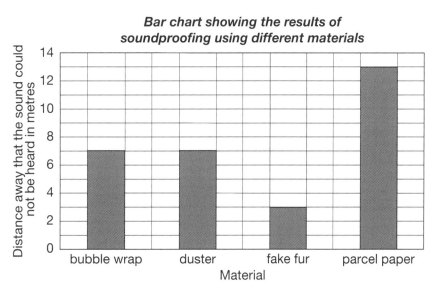

Bar chart showing the results of soundproofing using different materials

Distance away that the sound could not be heard in metres

Material: bubble wrap, duster, fake fur, parcel paper

a What things do Adya and Dembe need to keep the same to make this a fair test?

....

b At what distance could Adya and Dembe no longer hear the buzzer when it was

wrapped in a duster? ....

c Which was the best soundproofing material? ....

d Explain why you chose this material. ....

....

e Predict which of these materials is best at soundproofing. Circle one.

| carpet | kitchen foil | plastic bag | wrapping paper |

Musical instruments

A musical instrument can make sounds of different pitch by playing it in some way. The pitch is due to the speed at which part of the instrument **vibrates**. When someone is playing the instrument they are changing part of it to make different-pitched sounds that we call musical notes.

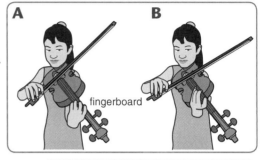

In the violin the strings have different thicknesses, which makes them vibrate at different speeds and produce notes of different pitch. The violinist can change the pitch of each string by pressing it down at different places on the fingerboard. This changes the length of the string. A shorter length makes a higher pitch than a longer length. She can also raise the pitch by tightening the string.

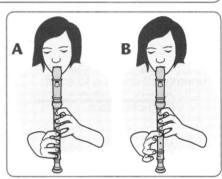

When you play a recorder, the air inside it starts to vibrate, and this vibrating air produces a note. If you cover all the holes on the recorder with your fingers, you play the lowest note on the instrument because all the air inside it is vibrating. If you lift up some of your fingers, the note becomes higher. This is because you have made the length of air vibrating inside the instrument shorter.

Key words

- musical notes
- violin
- recorder
- vibration
- pitch
- instrument

Have a go!

a Look at the two violinists above.

Which one is playing the higher pitched note? ☐

b Explain your answer. ....

....

....

....

....

Science challenge

1 Look at the two recorder players opposite.

   a Which one has got the most air vibrating in the instrument? ☐

   b Which one is playing the higher pitched note? ☐

   c If the player moved all her fingers from over the holes, what would

      happen when she blew down the mouthpiece? ....

      ....

   d Explain your answer. ....

      ....

2 Joe plucks a string on his violin without holding his finger down. The
   whole length vibrates. He then puts his finger a third of the way along and
   plucks the other two thirds. He moves his finger back to a quarter of the
   way along and plucks the remaining three quarters of the string.

   a Describe the change in pitch when   b Describe the change in pitch when
      Joe plays the first two notes.   he plays the third note.

      ....

Scientific enquiry

1 Sing the following words and say whether they are low or high pitched
   sounds. Circle your answers.

A Weeeeeee	low high
B Wurrrrrrrrr	low high
C Herrrrrrrrr	low high
D Pie	low high
E Paw	low high
F Ping	low high

2 Spot high and low sounds in your name. For example, the author's name
   is Peter Riley: this breaks down to *Pe* (high) *ter* (low) *Ri* (high) *ley* (low).
   Write your name here, identify the high and low sounds in it and sing it.

   ....

Circuits of electricity

> ### Key facts
>
> The parts of a circuit are called the components. The component that generates **energy** from the chemicals it contains is called the cell (battery). The energy is used to push extremely tiny **particles** called **electrons** around the circuit. The flow of the electrons makes the current of electricity in the circuit. Cells (batteries) have two points on them called terminals. One is a positive terminal and the other is a negative terminal. When we switch the circuit on, the cell (battery) pushes the electrons from the negative terminal, through the other components, to the positive terminal.
>
> The wires are long thin pieces of metal, usually coated in plastic. The metal part is an **electrical conductor** and the plastic is an **electrical insulator**, which stops the current flowing out of the sides of the wire. You control the flow of the current by moving a switch.
>
> The lamp is a common component in circuits. A lamp has a wire inside it that gets so hot when an electric current passes through it that it glows and gives out light.
>
> Some circuits have a component called a buzzer, which has a **material** that **vibrates** when electricity passes through the component – the vibrations make a sound.
>
> An electric motor is a component that uses the energy in the current to make a shaft spin.
>
> The components must be wired up to each other firmly so that no gaps in the circuit develop when we switch it on. If there are gaps, the current will not flow.

> ### Key words
>
> - electrical conductor   • electrical insulator
> - switch   • cell (battery)   • components   • electrons

Have a go!

Match the component to the description by drawing lines between them.

A switch	**1** pushes the electricity through the wires
B lamp	**2** gives off light when electricity passes through it
C wire	**3** makes a noise when electricity passes through it
D cell (battery)	**4** used to break the flow of electricity, to turn the lamp on and off
E buzzer	**5** links together different components in a circuit

Science challenge

When scientists make drawings of electrical circuits they use symbols.
The symbol for a wire is a straight line. Here are some other symbols.

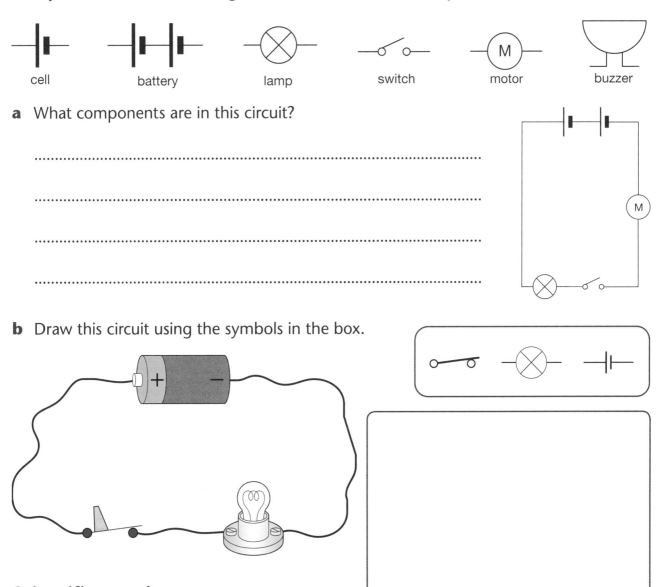

cell   battery   lamp   switch   motor   buzzer

a What components are in this circuit?

....

....

....

....

b Draw this circuit using the symbols in the box.

Scientific enquiry

Look at the picture below. If you wanted to make the lamp brighter,
which of these things could you do? Circle one.

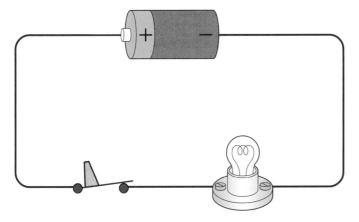

remove the switch

add another lamp

add another cell (battery)

remove the cell (battery)

add another wire

Investigating electricity

Key facts

Mains electricity is very powerful and we must never use it in investigations as it can kill people. The electricity cells (batteries) generate is much less powerful and is suitable for electrical investigations. We measure the power of electricity in **volts**, and the side of the cell (battery) displays its **voltage**. Most cells (batteries) we use in scientific investigations have a voltage of 1.5 V.

We can increase the power in an electrical circuit by joining two or more cells (batteries) together. We connect the negative terminal of one cell (battery) to the positive terminal of the other. If we connected two similar terminals together, each would push **electrons** towards the other and no current would flow.

When a **conductor** allows electricity to pass through it, some of its **material** tries to stop the flow of the current. This **property** is called resistance. The wire we use in lamps has so much resistance that some of the **energy** in the electric current turns to heat and light, and the wire glows and makes the lamp shine. The wires we use in other parts of the circuit have much less resistance and allow the current to flow more easily. However, as a long wire has more material in it than a short wire it also has greater resistance.

We can think of the flow of electrical current in wires like the flow of water in a river or pipe. A wide river or thick wire offers a low resistance to the flow of water or electricity while a narrow pipe or thin wire offers a much higher resistance to flow.

Key words

- volt
- voltage
- resistance
- conductor
- positive terminal
- lamp
- wires
- negative terminal

Have a go!

Shivi wants to connect two cells (batteries) to provide extra power in a circuit. Circle the correct way to connect them to produce a current.

> negative to negative terminals
>
> positive to negative terminals
>
> negative to negative terminals

Science challenge

Jack investigated how using different thicknesses of wire altered the brightness of the lamp. The table shows his results.

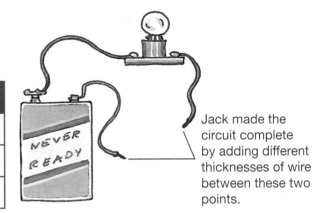

Jack made the circuit complete by adding different thicknesses of wire between these two points.

Thickness of wire	Brightness of lamp
thin	dimmer than normal
thick	normal
very thick	brighter than normal

a Describe what Jack's results show. ....

....

b Explain why Jack got these results. ....

....

c Describe how Jack could make this a fair test. ....

d Use the symbols in the box to draw a circuit diagram to represent the circuit that Jack used.

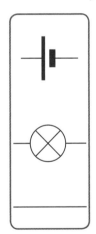

Scientific enquiry

a Look at the diagram you drew in answer to part **d** of the **Science challenge** and explain how you could use it to see if different objects were made of materials that were conductors or insulators.

b Describe how you could use the same circuit to investigate the effect of wire length on the flow of electricity in the circuit.

c What results do you predict?

How much can you remember? Test 3

1 **(a)** What unit do we use to measure force?

.... [1]

(b) What are the instruments we use for measuring force in the two pictures?

.... [1]

(c) What are the instruments measuring? Tick (✓) one box.

mass ☐   volume ☐   weight ☐ [1]

(d) What readings on the scales do the potato and melon show?

potato ....   melon .... [2]

2 **(a)** Look at these materials. Circle the one that things will slide on best.

glass   rubber mat   pavement [1]

(b) What is the force that acts when two surfaces rub together? .... [1]

(c) What is the force that acts on an object when it moves through the air?

.... [1]

3 **(a)** What are the two poles of a magnet called? .... [2]

(b) Ravi has two magnets, ten cards and a steel paper clip. He wants to compare the strength of one of the poles of the two magnets.

Here are sentences he wrote for his investigation, but they are in the wrong order. Write a number in each box to put the sentences in the correct order.

Add more cards until paper clip falls away. ☐

Use cards and paper clip on second magnet. ☐

Put clip on card and see if magnet holds it. ☐

Put a card next to a pole of the first magnet. ☐ [3]

(c) Describe how Ravi will know which magnetic pole is the stronger.

.... [1]

4 (a) Draw lines to match these words to their meanings.

A light source		1 something that does not let light pass through it
B light ray		2 when light bounces off something
C opaque		3 a line we draw with an arrow to show how light travels
D reflection		4 something that makes light

[4]

(b) Draw a beam of light with direction arrows on the picture to show how the mouse can see the cheese.

[2]

5 Ella plays a note on her guitar. She decides she needs to make the string produce a higher pitched note.

State two ways she could do this.

1 .... [1]

2 .... [1]

6 (a) Name the symbols used in this circuit diagram.

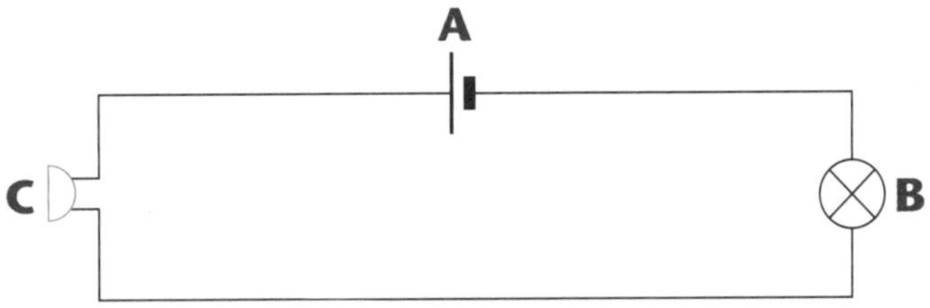

A ....   B ....   C .... [3]

(b) Write down what each component does. [3]

Glossary

Abdomen The lower part of the body below the chest.

Absorbent Able to take up a liquid.

Adapt / adaptation When an organism is suited to where it lives, it is said to be adapted to that habitat.

Air resistance A type of friction caused by air rubbing against the surface of something moving through it.

Atom A very tiny particle from which all substances are made. There are many different atoms making many different substances. For example, carbon atoms can make diamonds.

Axis An imaginary line that runs through the Earth from pole to pole.

Balanced diet A diet that contains the correct amounts of nutrients to keep you healthy.

Boiling The condition in a liquid when it is as hot as it can get and is turning into a gas as fast as it can.

Carbohydrates Foods such as pasta and rice, which are rich in energy that the body can use quickly.

Carnivore An animal that eats other animals and does not eat plants.

Classification The grouping of living things according to the features of their bodies.

Condense / condensation The turning of a gas into a liquid. Also the name of a liquid that has condensed on a cold surface.

Conductor A material through which a current of electricity can pass easily. Some materials also allow heat to pass through them easily.

Conserve / conservation Protecting the habitats of plants and animals so that they will not become extinct.

Consumer An animal that eats other living things (either plant or animal).

Decibel A unit for measuring the loudness of sound.

Dissolve A process in which a substance breaks down into tiny particles in a liquid and spreads out in it.

Electrical conductor A material through which a current of electricity can pass easily.

Electrical insulator A material through which a current of electricity cannot pass easily.

Electrons Very tiny particles that carry the electric current round a circuit.

Energy Something that has different forms such as light, heat and movement. It gives things power to move and living things power to carry out all the life processes.

Evaporate / evaporation The process of turning from a liquid into a gas.

Extinction The dying out of all the individuals of one kind of plant or animal so that none are left alive.

Factors Things that can affect something. For example, light, warmth and the amount of water are all factors that can affect the growth of a plant.

Fair test A test in which the scientist keeps all the factors the same except one. The scientist varies this factor to see how it affects the activity under investigation. For example, in a fair test to see how temperature affects plant growth, the plants all receive the same amount of light and the same amount of water but are kept at different temperatures.

Fats Substances found in foods such as butter, which give you a store of energy.

Fertilisation The process in which part of a pollen grain joins with part of an ovule in the ovary to make a new plant in a seed.

Fertilisers Substances containing a lot of minerals, which plants take up in water through their roots.

Fibre A substance in some foods that helps food to be moved through the digestive system as digestion takes place.

Filter Something with very small holes used in filtering. A filter is often made using filter paper inside a funnel.

Filtrate A substance that passes through a filter.

Food chain A diagram showing a line of living things (usually a plant and one or more animals) linked by arrows pointing from the living thing being eaten to the living thing that is eating it (from food to feeder).

Force A push or a pull.

Fossil fuels Substances like oil, natural gas and coal formed from the decomposition of the bodies of living things that lived millions of years ago.

Freeze The process in which a liquid turns into a solid at a certain temperature (the freezing point).

Friction A force that acts when two surfaces slide over each other.

Fuel A substance that is burnt to provide heat or energy to make machines such as cars move.

Gas A state of matter in which the matter flows easily, can change shape easily, and can change volume easily (can be squashed into a smaller space).

Germinate / germination The process in which a root and then a shoot emerge from a seed.

Gravity A force between any two objects in the universe. On Earth it is the force that pulls all objects towards the planet's centre, resulting in things falling to its surface and down holes.

Greenhouse gas A gas in the atmosphere that holds on to heat from the Sun and makes the atmosphere warmer.

Habitat A place where a plant or animal lives. It shares this place with other plants and animals. Examples of habitats are forests, fields, mountains, deserts, swamps, ponds, rivers, the seashore, the open sea and the ocean floor.

Herbivore An animal that eats only plants or parts of plants. It does not eat animals.

Glossary

Horizon The line between land in the distance and the sky. It may be slightly curved, like the horizon between the sea and sky, or have the outline of buildings on it, like the horizon in a city.

Irreversible change A non-reversible change in which the substances produced cannot be changed back to the original substances easily. For example, when a material is burnt.

Light source Something that gives out light.

Limbs Extensions of the main body of an animal (the arms and legs), which help the body to move.

Liquid The state of matter in which the matter flows, can change shape easily, but cannot change volume (cannot be squashed into a smaller space).

Manure Solid waste from animals, usually cattle, which we use as a fertiliser for the soil. It also improves the texture of the soil, which helps plant growth.

Mass The amount of matter in a substance.

Material Any solid, liquid or gas that is composed of matter.

Matter The 'stuff' from which things are made – atoms and groups of atoms called molecules.

Minerals This word has three meanings.
1 The nutrients in the soil that plants take up by their roots. **2** Nutrients in food that keep us healthy and help us grow.
3 Substances, often made of crystals, which join together to form rocks, precious stones and metal ores such as iron ore.

Molecule A group of atoms joined together.

Muscle A fleshy part of the body that can contract and bring about movement.

Natural gas A gas trapped in rocks, which has formed from the decomposition of dead bodies of living things that lived millions of years ago.

Nerves Long, thin, thread-like structures in the body that carry messages between the body parts in the form of tiny electrical currents.

Nutrition The processes involved in a living thing acquiring food and water. Animals feed and drink to take in food and water. Plants make their own food using sunlight and draw in water and minerals through their roots.

Opaque A property of a material that does not let light pass through the material.

Particles This word has two meanings.
1 Very small pieces that you can see, such as grains of sand and dust. **2** Extremely small structures called atoms, which we can only see using very powerful microscopes. We also use the word to describe electrons, which are even smaller than atoms and form part of their structure.

Permeable Something that will let other substances like water pass through it.

Poles This word has two meanings in Science. **1** The ends of the Earth's axis in the northern and southern hemispheres. **2** Regions at the ends of a bar magnet, which are attracted to the north and south poles of the Earth.

Properties Features of a material such as its hardness, smoothness and permeability to water.

Proteins Foods that help the body grow and repair injuries.

Reaction A word scientists use to note that a change has taken place. When a candle burns, a chemical change takes place in the wax. We can call this a chemical reaction.

Reversible change A change in which the substance produced can be changed back to the original substance easily. For example, liquid water can be frozen to make solid ice but can be melted to make water again.

Solid The state of matter in which the matter does not flow and so cannot change shape or volume (cannot be squashed into a smaller space).

Solute A substance that dissolves in a liquid (solvent) to form a solution.

Solution A liquid (solvent) that has one or more substances (solutes) dissolved in it.

Solvent A liquid in which a substance (solute) will dissolve to form a solution.

Sound waves Waves of energy that can pass through solids, liquids and gases and on reaching the ear produce the sensation of hearing.

Streamlined A description of the shape of an object that has low air resistance so the object can move quickly. Racing cars and aeroplanes have streamlined shapes.

Substance A general name for any kind of matter, whether it is solid, liquid or gas.

Temperature A measure of the hotness or coldness of something. Temperature is measured in °C.

Thorax The part of the body between the head and the abdomen. In the human body it is commonly known as the chest.

Torso The part of the human body without the head or limbs. It is made from the thorax (chest) and abdomen.

Transparent A property of a material that lets light pass through the material.

Vibrate A movement in which something moves backwards and forwards or up and down very quickly.

Volt / voltage The volt is a unit for measuring the power of a cell (battery) to send a current around a circuit. The symbol for the unit is V. The voltage is the value in volts shown on the side of a cell (battery).

Volume The amount of space that something takes up. We measure volume in units such as cm³.

Water vapour Water in the form of a gas in the air.

Waterproof A property of a material that prevents the material from letting water pass through it.

Weight A measure of the force of gravity pulling down on something.

Wilt The drooping of plant due to a lack of water to support it.

Extracts adapted with permission of Mark Levesley.

Note: Whilst every effort has been made to carefully check the instructions for practical work described in this book, schools should conduct their own risk assessments in accordance with local health and safety requirements.

Every effort has been made to trace all copyright holders, but if any have been inadvertently overlooked the Publishers will be pleased to make the necessary arrangements at the first opportunity.

Hachette UK's policy is to use papers that are natural, renewable and recyclable products and made from wood grown in sustainable forests. The logging and manufacturing processes are expected to conform to the environmental regulations of the country of origin.

Orders: please contact Bookpoint Ltd, 130 Milton Park, Abingdon, Oxon OX14 4SB. Telephone: (44) 01235 827720. Fax: (44) 01235 400454. Lines are open 9.00–5.00, Monday to Saturday, with a 24-hour message answering service. Visit our website at www.hoddereducation.com.

© Peter D. Riley 2013
First published in 2013 by
Hodder Education,
An Hachette UK Company
Carmelite House, 50 Victoria Embankment,
London EC4Y 0DZ

Impression number   7 6 5 4
Year   2020 2019 2018 2017

Cover illustration by Peter Lubach
Illustrations by Planman Technologies, Barking Dog and Pantek Media Ltd
Typeset in ITC Stone Sans Medium 12.5/15.5 by Planman Technologies
Printed in India

A catalogue record for this title is available from the British Library.

ISBN: 978 1444 178302